nanodesign

SOME BASIC
QUESTIONS

nanodesign

SOME BASIC
QUESTIONS

Wolfram Schommers
Karlsruhe Institut für Technologie (KIT), Germany

 **World Scientific**

NEW JERSEY · LONDON · SINGAPORE · BEIJING · SHANGHAI · HONG KONG · TAIPEI · CHENNAI

Published by

World Scientific Publishing Co. Pte. Ltd.

5 Toh Tuck Link, Singapore 596224

USA office: 27 Warren Street, Suite 401-402, Hackensack, NJ 07601

UK office: 57 Shelton Street, Covent Garden, London WC2H 9HE

Library of Congress Cataloging-in-Publication Data
Schommers, W. (Wolfram), 1941– author.
    Nanodesign : some basic questions / Wolfram Schommers, Karlsruhe Institut für Technologie (KIT), Germany.
        pages cm
    ISBN 978-9814520348 (hardcover : alk. paper)
    1. Nanotechnology. I. Title.
    T174.7.S36 2014
    620'.5--dc23

                                                                                    2013035517

British Library Cataloguing-in-Publication Data
A catalogue record for this book is available from the British Library.

In-house Editor: Rhaimie Wahap

Typeset by Stallion Press
Email: enquiries@stallionpress.com

Printed in Singapore

PREFACE

■ ■ ■

In nanoscience and nanotechnology we move between two extreme poles that are expressed by the two extreme states "infinite life" and "total destruction." On the one hand, through nanotechnology, aging might soon be a fact of the past; on the other hand, in the nano realm uncontrolled processes could take place, leading to total destruction of the living conditions on the earth.

The reason for this situation is obvious: in nanoscience we are able to manipulate and to change the material world at the ultimate level; this is the level at which the properties of matter emerge and biological individuality (humans and animals) comes into existence.

Nanoscience will open up completely new perspectives on all scientific and technological disciplines. But, as we will discuss in this monograph, there will not only be positive perspectives in this field but also threats.

Without doubt, the threats in nanotechnology are large–larger than those we know in connection with nuclear energy. As said, basically we always take a position between 'infinite life' and 'total destruction.' Thus, the following point is of particular relevance: In order to avoid undesirable developments, the theoretical (computational) analysis of certain processes is not only desirable but absolutely necessary, i.e. we have to analyze the situation *before* we start the production of certain nanosystems. Question: Are the present physical laws sufficient for such a preview? Is our scientific "world view" adequate for the understanding of nanotechnological manipulations, just as when we work in the realm of brain research? These are in fact basic questions that have to be answered when we enter the field of nanodesign.

•

During the last 150 years, modern technology has developed fantastically. We have designed and produced cars, planes, and a lot of other interesting and useful things. For 50 years, microtechnology has been a factor. This discipline generates its features near 1 micrometer (one-millionth of a meter). On this strongly reduced scale mechanical and electronic devices could be constructed and realized by sophisticated production methods. A prominent example is the microscopic transistor; single chips could be formed with a large number of such transistors and, in this way, a lot of information could be stored. However, all these things are nothing when we enter the field of nanotechnology, i.e. when we compare the products of microtechnology with those of nanotechnology.

In nanotechnology we work at the atomic level, and this discipline generates its features near 1 nanometer (one-thousandth of a micrometer). While the chips of microtechnology can store a lot of memory, a nanodevice can store an amount of information that is really unimaginable and goes far beyond microtechnology. For instance, the usual credit card would be able to store nanotechnologically a movie with the complexity of *Ben Hur*, which would run a good many years without any interruption. For another example, let us consider a mechanical nanodevice with a rotating unit. This unit could rotate with 10^9 revolutions per second. This is really a very large number and is unimaginable. For comparison, if the wheels of a car could rotate with such a large number of revolutions, the car would circle the earth more than 80 times in a second. In Chap. 2 we will discuss such a nanosystem; it is a very small electrical nanogenerator, with an extension of 20 nm. A hair has a diameter of approximately 80,000 nm; in other words, the diameter of the hair is 4000 times larger than the size of the nanogenerator. This is almost unimaginable.

•

In conclusion, nanosystems are unimaginably small and very fast. But this is not the essential point in connection with nanoscience and nanotechnology. The essential point is that we work here at the "ultimate level." What does the term "ultimate level" mean? This is the *lowest* level at which the properties of our world emerge, at which functional matter can exist, at which biological individuality comes into existence.

This situation can be defined in absolute terms: *"This is not only the strongest material ever made, this is the strongest material it will ever be possible to make"* (D. Ratner and M. Ratner, *Nanotechnology and Homeland Security*).

In fact, we learned to manipulate matter at its ultimate level on the basis of specific experimental devices. For the first time, single atoms could be moved in a controlled manner from one position to another, and nanoscience (nanotechnology) became an important scientific and technological discipline.

Since we work at the ultimate level, we have to apply the basic laws of theoretical physics. This is an important point and has to be considered very carefully.

•

Again, in nanotechnology we consider the world at its ultimate level, theoretically as well as experimentally. As has already been remarked, this is exactly also the level at which 'biological individuality' comes into existence, at which life has its basis. Is the present formulation of theoretical physics adequate for the understanding of the facts at the ultimate level at which nanotechnology (nanoscience) works? This question will be discussed extensively in this book.

Working at the ultimate level means in particular that we do not change the features of a biological system only superficially during the nanoexperimental impact, but the "heart" of the matter (system) is directly concerned, i.e. its basic characteristics. In other words, in general we not only change a certain system but a completely new system will be created. This can be dangerous and we have to be careful.

The creation of a new system normally takes place at the nanolevel by "self-organizing processes." We put a certain (biological) system — say, A — into a certain environment, and system A develops in the course of time toward a new system — say B — and this process takes place without any activity of the nanoengineer.

If these 'self-organizing processes' are not under control it might lead to a disaster, since we do not know the structure of system B at the beginning, when we start the process. Let us give an example that has been discussed in the literature: system B could be a self-replicating nanorobot that transforms everything on the earth into a differentiationless mass. In other words, in such a case we have 'total destruction'; biological individuality and life, respectively, would no longer be possible on the earth. This might be considered as exaggerated. But it is not. If there is the slightest possibility for the appearance of such self-organizing processes, we must stop it, i.e. a start must be avoided; this is the worst case and must be avoided.

We must have reliable theoretical models for the description of any kind of self-organizing processes and, as has already been remarked above, these models have to be applied before we start the process experimentally. Are our theories realistic enough for such sensible situations?

On the other hand, due to nanotechnology there will be a lot of positive facts. Cancers will be cured, along with most other ills of the flesh. Aging, or even routine death itself, might become a thing of the past.

In conclusion, in nanoscience (nanotechnology) we incessantly take a state between two limiting cases, i.e. 'total destruction' and 'infinite life'. Can a human being psychologically master such a situation?

•

Brain functions can be influenced by nanotechnological manipulations. Certain experiments with animals have already been done, and they suggest that man's intelligence could also be increased considerably. Scientists added a certain gene to the brain of a mouse and obviously improved the efficiency of the brain functions; similar experiments are explicitly planned for man. In fact, it has been prognosticated that it will take approximately 30–50 years to develop nanotechnological means for the creation of superhuman intelligence. In conclusion, it is assumed that we will learn to increase our own effective intelligence or to cure Alzheimer's disease.

Are the theoretical tools developed so far sufficient for understanding this kind of nanotechnological manipulation? Are we really in a position to describe biological phenomena? We will discuss this point critically by means of basic principles. Let us mention here some important facts.

Two features are obviously essential for the perception of the world outside by man:

(1) The "strategy of nature" is important, and it is dictated by the principles of evolution;
(2) How do other biological systems (animals) perceive the world?

Evolution reflects pragmatic features. The perception of true reality in the sense of precise reproduction implies that with growing fine structure in the picture, increasing information of actual reality outside is needed. Then, evolution would have developed sense organs with the property of transmitting as much information from reality outside as possible. But the opposite is the case: the strategy of nature is to take up as little information from the outside world as possible.

This in particular means that there cannot be a one-to-one correspondence between what we have in front of us in everyday life (it is the inside world) and what is actually outside. Reality outside is not assessed by "true" and "untrue" but by "favorable to life" and "hostile to life." This is probably the most important idea with respect to the phenomenon of evolution.

This is exactly also the case for animals. Here, too, "favorable to life" is the most important factor. In fact, certain experiments within the frame of behavior research confirm that.

In summary, from the strategy of nature and the behavior of animals it follows that biological systems (for example a human being) never have the basic, objective reality in front of and around them, but it is in any case a species-dependent reality. (This point will also be discussed in detail in this monograph.)

Such species-dependent realities are essentially influenced by the brain functions of the biological system. The structure of this world in front of a species is of particular relevance to survival. If this "world view" is disturbed, the system will possibly have problems with survival.

We may increase the intelligence through nanotechnological operations. However, since all brain functions are more or less interrelated, the change of one brain function will have consequences for the others, and this can lead to serious problems. For example, those brain functions could be negatively changed that are responsible for the correct construction of the world view of the biological system. Then, we disturb the relationship between the system and its environment, and this mismatch can be large, with the effect that the system (for example a human being) becomes unable to cope with life.

In other words, we have to be careful when we try to change nanotechnologically certain brain functions. If we change one brain function positively, other brain functions are influenced automatically and simultaneously, and we do not know how. We may state quite generally that changes of brain functions can lead to serious problems. Before we start such experiments, we have to know precisely how the brain works. This can only be assessed on the basis of the fundamental laws of physics. Again, is the present theoretical framework of physics reliable enough for such enterprises? The scientific community has to investigate this point critically. Moreover, in order to avoid a mismatch, the world view of physics has to be identical with that of the biological system that is produced by the brain by means of the outside information.

•

In this monograph, some principal remarks on the construction of nanosystems are made. Self-organizing processes in connection with the basic laws of theoretical physics are analyzed. The present world view, on which present physics is based, has been discussed critically. The strategy of nature in the design of life and some basic results of behavior research are included.

This book is based on a lecture that I have given within the "Distinguished Speaker Series 2011–2012" at the University of Texas at Arlington. I am grateful to all my colleagues and students at Arlington for fruitful discussions.

Wolfram Schommers
Arlington, Texas, USA
Karlsruhe, Germany

CONTENTS

■ ■ ■

Chapter One

NANODESIGN: DESCRIPTION AT THE ULTIMATE LEVEL

■ ■ ■

There is no doubt that nanoscience will be the dominant direction for technology in this century, and this science will influence our lives to an extent impossible in years past: specific manipulations of matter will open up completely new perspectives on all scientific and technological disciplines. To be able to produce optimal nanosystems with tailor-made properties, it is necessary to analyze and construct such systems in advance by adequate theoretical and computational methods.

But, as is well known, there are also threats connected with nanotechnology, specifically with respect to biological systems. For example, self-assembly can become an uncontrolled process. To avoid undesirable developments, the theoretical (computational) analysis of such processes is not only desirable but also absolutely necessary. In conclusion, threats can extensively be eliminated by systematic theoretical analyses of the system under investigation using adequate laws from theoretical physics.

Thus, the computational and theoretical methods of nanoscience are not only essential for the prediction of new and custom nanosystems but also help to keep nanotechnology under control. In this book, we will not discuss all these methods in detail, but rather the general features and the basics of them which are dictated by the necessities at the nanolevel.

1.1 TREATMENT ON THE SAME FOOTING

In nanoscience and nanotechnology we work at the ultimate level, where the properties of matter emerge and where, in particular, biological individuality comes into existence. This is of fundamental relevance, especially to the development of theoretical and computational methods that are the basis for the understanding of nanophenomena, and of course also to further technological developments.

In traditional technologies (micro- and macrotechniques), engineers do not work at the ultimate level. They use more or less phenomenological descriptions, which, in general, cannot be deduced from the basic physical laws; each discipline has its own description.

In nanoscience and nanotechnology we have various systems and disciplines: materials science, functional nanomaterials, nanoparticles, food chemistry, medicine with brain research, quantum and molecular computing, bioinformatics, magnetic nanostructures, nano-optics, nanoelectronics, etc. Although we have here a lot of directions and disciplines, respectively, which are partly very different in character, we are working here on the same theoretical footing — in contrast to the traditional technologies. In nanotechnology we have one theory for all these disciplines and all nanophenomena, and this is given by the basic laws of theoretical physics. That has at least two consequences:

(1) Since we are working here at the ultimate level, the properties of matter and functional matter are defined in absolute terms. What does "ultimate level" mean? At this level the properties, for example those of a material, cannot be further improved. We have reached the highest level. It is the level at which the properties emerge. At this level also biological individuality comes into existence, where life has its origin.

(2) Working at the ultimate level also means that any change in the basic physical laws will directly influence nanoscience and nanotechnology, without any intermediate step. This can lead to completely new perspectives in connection with applications. Therefore, to work on nanoscience also means to develop the basic

laws further, if there is a necessity for that. This could be important just in connection with quantum theory. New experimental results indicate that there is in fact a need for that.

In connection with brain research, we must even go a step further, because here the world view is of particular relevance. The theoretical world view must be in accord with the world view, which the brain develops for the world outside unconsciously. This topic will also be discussed in detail in this book.

1.2 COMPLEXITY OF NANOSYSTEMS

Nanosystems are interesting and behave in relevant cases quite differently from systems used in micro- and macrotechnology. However, many researchers discuss nanosystems on the basis of traditional thinking. Why? They very often study the effects by means of static building blocks, as we do in connection with microsystems. Normally it is assumed that the properties of nanosystems (for example a nanocluster) are due to the relatively large surface area; the relative number of surface atoms increases with decreasing size. This effect modifies the structure (in comparison with that in the bulk) of the building blocks but they remain static elements within this view. However, this structure aspect is only one typical effect in connection with nanosystems. There is another important point: nanosystems behave strongly anharmonically, much more than in micro- and macrosystems, i.e. the dynamics is also modified, which may have the effect that the whole nanosystem can transform spontaneously and there can be many coexisting structural states. As a typical example, let us mention some interesting features in connection with nanoclusters.

Nanoclusters can be in an excited state (like atoms). After a certain time the excited cluster transforms spontaneously to the ground state without external influence. There can be more than one ground state and it is quite a matter of chance to what ground state the cluster transforms from the excited state. The various ground states may differ in both the inner structure and the outer shape.

In other words, a free cluster or a cluster on a surface is for a certain time interval a static system. But suddenly it transforms spontaneously to another system with another inner structure and another outer shape. In micro- and macrotechniques we do not observe such effects; here we have for all times one structure and one shape.

We may conclude that the behavior of systems, relevant in nanotechnology, is in general complex and is obviously not comparable with the behavior of micro- and macrosystems. For the description of such nanosystems the theoretical and computational tools have to be selected very carefully. In this connection the interaction laws (potentials) between the atoms forming a nanosystem are critical functions, because the structure and dynamics of such systems are very sensitive to small variations in the potentials. We will discuss this point in Chap. 2.

1.3 NANOTECHNOLOGICAL CHANGES

It is generally accepted that nanobiology and nanomedicine will influence our future considerably. Here also, theoretical and computational nanotechnology will play an important role. What mathematical tools do we need here? Are there principal limitations in connection with the understanding of biological systems? The latter question directly arises when we consider the human brain as a mathematical object; a physical view is always connected with a mathematical framework. Is a complete description of the human brain possible on the basis of the laws of theoretical physics? In the next section and particularly in Chap. 2, we will discuss this problematic point in more detail.

The impact of nanotechnology will be tremendous. The manipulation of atoms, molecules, etc. will allow us to construct new technological worlds and will bring fundamentally new possibilities in the field of medicine. Just in the case of nanobiotechnology, big changes are expected already in the near future. It has been speculated that through nanotechnology our bodies will be transformed into undecaying systems of infinite life. It is particularly planned to change the brain functions in order to increase man's intelligence.

On the other hand, the threats in nanotechnological manipulation are large, just in connection with self-organizing processes (self-assembly), and such processes can lead to uncontrolled processes and strong destructions may appear. It has been speculated that even the entire earth could be transformed into a system hostile to life. In other words, biological individuality, i.e. life, would no longer be possible on the earth.

As has already been remarked, within nanotechnology and nanoscience we move between two limiting cases, between two poles, and these poles are given by total destruction and infinite life. In Fig. 1 this situation is summarized.

The probability that one of these limiting scenarios becomes reality increases in the course of time; it increases with the number (n) of nanotechnological manipulations. We know that n does not remain constant or behaves linearly, but rather such a number of developments will increase exponentially. Figure 2 shows this in plain terms. In summary, the probability of approaching one of the limiting cases (Fig. 1) will grow considerably in the course of time.

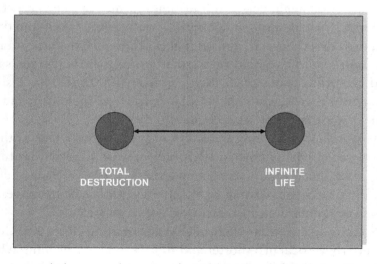

Fig. 1 In nanotechnology we move between two poles: total destruction and infinite life. It must be a challenge for science to understand all these processes theoretically in order to avoid disasters.

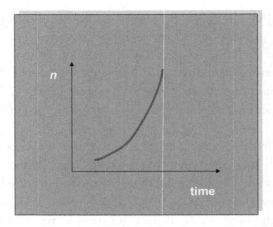

Fig. 2   The probability for the two limiting cases, total destruction and infinite life, increases in the course of time, because the number (*n*) of nanotechnological activities increases in the course of time. The growth of this curve can be assumed to be exponential.

Viewing the growth of nanotechnology in the course of time like an exponential curve leads to the effect that the future rushes to us. The more the exponential curve growth, "the larger is each subsequent bound upward. It takes a long time to double the original value, but the same period again gets you four times farther up the curve, then eight times ... so that just ten doublings, you've risen a thousand times as far, then two thousand, and on it goes. Note this: the time it takes to go from one to two, and then from two to four, is just the same period needed to take that mighty leap from 1000 to 2000. A short time later we're talking a millionfold increase to a single step, and the very next step after that is two millionfold ... "

"History's slowly rising trajectory of progress over tens of thousands of years, having taken a swift turn upward in recent centuries and decades, quickly roars straight up some time after 2030 and before 2100. That's a Spike. Change in technology and medicine moves off the scale of standard measurements: it goes asymptotic ... "

"So the curve of technological change is getting closer and closer to the utterly vertical in a shorter and shorter time. At the limit, which is reached quite quickly (disproving Zeno's ancient paradox about the tortoise beating Achilles if it has a head start), the curve tends toward infinity. It rips through the top of the graph and is never seen again. At the Spike we can confidently expect that some form of intelligence (human, silicon, or a blend of the two) will emerge at a posthuman level. At that point, all the standard rules and cultural projections go into the waste-paper basket."[1]

This accelerating world of drastic technological change is the reason why this posthuman phase will be reached relatively soon, and this will possibly have drastic consequences. In his book *The Spike*, Damien Broderick formulates that as follows: "We can expect extraordinary disruptions within the next half century. Many of these changes will probably start to impact well before that. By the end of the twenty-first century, there might well be no humans (as we recognize ourselves) left on the planet — but, paradoxically, nobody alive then will complain about that, any more than we now bewail the loss of Neanderthals."[1]

Such a situation in connection with an accelerating world of drastic technological change is without any doubt dangerous and can lead to an uncontrolled situation. Within such a situation the probability of reaching one of the poles (total destruction and infinite life) in Fig. 2 increases enormously. Therefore, each step in relevant nanotechnological developments and changes has to be taken carefully in order to keep (nano)technology under control. The outcome has to be known *before* we start the process experimentally. Again, are the present theoretical and computational methods reliable enough to achieve that?

1.4 BRAIN FUNCTIONS

It is known that brain functions can be influenced by manipulation in the nano realm. Certain experiments with animals have already been done, and many scientists are firmly convinced that man's intelligence can also be increased considerably. In addition, is expected that serious diseases, for example Alzheimer's disease, can be cured in the future.

Are the theoretical tools developed so far sufficient for understanding all that? Are we really in a position to adequately describe

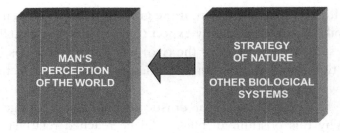

Fig. 3   On what factors is man's perception dependent? There are particularly two points: (1) the "strategy of nature" (principles of evolution); (2) we can also read our possibilities for perception from the behavior of other biological systems (animals).

biological phenomena? We will discuss this point critically, by means of basic principles.

1.4.1 World Views

What is essential to know about man's brain? How is the recognition system of man constructed? What is man's basis when he tries to judge the function of biological systems? Here two features are important (see also Fig. 3):

(1) The "strategy of nature," which is dictated by the principles of evolution;
(2) How do other biological systems (animals) perceive the world?

The strategy of nature, reflected by the principles of evolution, is pragmatic in character. It is the strategy to take up as little information from the outside world as possible. Reality outside is not assessed by "true" and "untrue" but by "favorable to life" and "hostile to life." This point is often underestimated and is presently not an object in the formulation of the physical laws.

Specific experiments within the framework of behavior research show that also in the case of animals the factor "favorable to life" is essential.

In summary, from the strategy of nature and the behavior of animals it follows that biological systems (for example a human being) never have the basic, objective reality in front of and around them. It is in any case a species-dependent reality. It is the world view that is used in everyday life. (This point will be extensively discussed in this monograph.)

These species-dependent realities of the various biological systems are essentially influenced by the brain functions. This structure of the world in front of and around a species, its world view, is of particular relevance to survival. If this world view is disturbed, the system will possibly have problems with survival.

1.4.2 Gödel's Theorem

There is another principal point. Understanding the brain by physical tools and models, respectively, means that the physical description of the brain must be based on a mathematical structure. Is that possible at all for a biological system that tries to make statements about itself?

Then, the question "What is the world view of an individual and how is it described?" is equivalent to the question "Can a mathematical system make statements about itself?" and this is because the individual wants to make statements about itself. It is an attempt at a self-related analysis. Here Gödel's theorem[2] comes into play: Let us assume that we have a consistent mathematical system. In accordance with Gödel's theorem, there will be true statements which are not included within the system; there are statements which cannot be proven to be true or wrong with system-specific axioms and rules. In other words, the system is not complete. A self-related analysis is not completely possible. So, we have to conclude that a human cannot completely understand his own brain. Or, more generally, a biological system is principally not able to make complete statements about itself. We know a lot about the brain. The problem is that we do not know what we are not knowing about it.

1.5 THE RELEVANCE OF BASIC QUANTUM THEORY TO NANOTECHNOLOGY

1.5.1 Temperature Effects

We have already mentioned that nanosystems behave in relevant cases quite differently from systems used in micro- and macrotechnology.

The main reason is the fact that the atomic (molecular) structure and dynamics of nanosystems are strongly dependent on temperature, much more than the systems of micro- and macrosystems. The reason for this temperature sensitivity is simple: with decreasing size the relative number of particles at the surface increases, and there are a lot of surface particles in the case of the usual nanosystems. Since the surface atoms are less bonded than the atoms (molecules) in the bulk, we expect that the structural and dynamical properties of nanosystems vary strongly with temperature.

Such theoretical and computational methods have to be chosen that are able to treat the anharmonicities realistically. For example, the molecular dynamics method fulfills this condition; in fact, it is able to treat anharmonicities without approximation. In Chap. 2, we will talk about this and other computational methods.

The molecular dynamics method is a classical description of many-particle systems, which is based on Newton's mechanics. However, within such calculations we need as input the interaction potentials between the atoms (molecules), and these interactions can only be determined quantum-theoretically.

1.5.2 Quantum Devices

It is not the goal to work under conditions where quantum effects are suppressed but to profit from them for the construction of new systems: electronic nanodevices, developments in nanomedicine and nanorobotics and drug design, quantum-electromechanical systems, quantum computers, etc. "In the last decade device fabrication and experimental control have progressed to such an extent that one can now see how quantum mechanics will be used to build a new technology."[3]

All these physical devices and systems operate on certain quantum principles. In Ref. 3, Milburn gave a list of such principles reflecting the key elements of quantum mechanics relevant to technological tasks: uncertainty principle, superposition, quantization (quantum size effect), tunneling, entanglement, decoherence. In this connection the following remark by Milburn is instructive: "A number of imperatives will drive the development of quantum technology. To begin, there is the quest for smaller and faster devices taking us to the nanoscale. At this scale quantum principles become manifest at low temperatures. Nanotechnology must take heed of quantum principles at some level. Any technology requires transducers and high precision measurement. Quantum theory has some very important things to say about measurement and its limits. Quantum technology will necessarily lead to new instruments. It is already clear that a number of communication challenges involving bandwidth and energy can be faced within a quantum context, with teleportation being the most surprising protocol. Finally there is the promise of quantum computing. In each case the quest to harness quantum mechanics for technological ends will bring new science along with new experimental opportunities."[3]

The basic laws of physics are without doubt of particular relevance to the development of nanotechnology. In Chap. 1, Sec. 1.1 we have stated the following: Working at the ultimate level also means that any change in the basic physical laws will directly influence nanotechnology, without any intermediate step. This can lead to completely new perspectives in connection with applications. Therefore, to work on nanoscience also means to develop the basic laws further. This could be important just in connection with quantum theory. New experimental results indicate that there is a need for that. We will discuss this point in more detail in the next section.

1.6 SUMMARY

(1) Nanoscience will be the dominant direction for technology in this century and probably beyond it. Specific manipulations of matter will open up completely new perspectives on all scientific and technological disciplines. Optimal nanosystems with tailor-made properties will be at the center and, for the construction of such systems, adequate theoretical and computational methods are necessary. But, due to the threats in connection with nanotechnology, particularly with respect to self-organization processes, the

theoretical (computational) analysis of such processes is not only desirable but absolutely necessary.

(2) In nanoscience and nanotechnology we work at the ultimate level, where the properties of matter emerge and where, in particular, biological individuality comes into existence.

Although we have in nanotechnology a lot of directions and disciplines, respectively, which are partly very different in character, we are working here on the same theoretical footing — in contrast to the traditional technologies. In nanotechnology we have "one theory" for all disciplines and all phenomena, and this is given by the basic laws of theoretical physics. Working at the ultimate level in particular means that any change in the basic physical laws will directly influence nanoscience and nanotechnology, without any intermediate step. This can lead to completely new perspectives in connection with applications.

In brain research, even the world view is of particular relevance. The theoretical world view (developed consciously) and the world view, which the brain develops unconsciously of the world outside, must be consistent.

(3) Nanosystems behave in relevant cases quite differently from systems used in micro- and macrotechnology. The whole nanosystem can transform spontaneously and there can be many coexisting structural states. As a typical example, we have mentioned some interesting features in connection with nanoclusters.

(4) Due to the possibilities in nanoscience, in future we will move between two limiting poles, which are given by total destruction and infinite life.

(5) From the principles of evolution and the behavior of animals, it follows that biological systems (human beings and animals) never have the basic, objective reality in front of and around them. It is in any case "only" a species-dependent reality. It is the world view that is used in everyday life.

These species-dependent realities, produced by the various biological systems, are essentially influenced by the brain functions. This structure of the world in front of and around a species, its world view, is of particular relevance to survival. Any disturbance of this world view can lead to basic problems; the system will possibly not have the chance to survive. All that has to be considered when we try to manipulate nanotechnologically the brain functions.

Chapter Two

PRINCIPAL REMARKS

■ ■ ■

In this monograph, some principal remarks on nanoscience and nanotechnology are made. What topics are relevant in this field? We will discuss some of the important features that might essentially determine the future direction of nanoscience and nanotechnology. We will not give a complete representation, but let us concentrate on a few specific facts that are relevant to the theoretical and computational treatment of nanosystems.

Here, some principal questions in connection with biological systems are relevant, but also with respect to nonbiological systems, particularly nanomachines and nanomaterials. In connection with nanomaterials the atomistic and electronic properties are of relevance; realistic determination of the structure and the dynamics of atoms (molecules) of such nanosystems is a particular challenge and needs a uniform basis.

In the study of such and similar systems, three points have to be considered carefully:

(1) What parameters are characteristic?
(2) What kind of physical laws have to be applied?
(3) What is the level of description?

Before we discuss these and further points, let us give three typical examples of nanosystems. The first example is from materials science, the second from food chemistry, and the third from biology;

we will in particular discuss the topic "medicine" in connection with brain research. Let us start with materials science; here we will restrict ourselves to nanoclusters. What are the properties of nanoclusters in comparison with a very large system (characterized by bulk states) which we experience in everyday life?

2.1 SIMPLE NANOCLUSTERS

Let us consider a simple cluster consisting of aluminum (Al) atoms. The cluster is made up of approximately 500 atoms, and it has a certain temperature. For this Alcluster, realistic model calculations have been performed. What does "realistic" mean here? Such small systems behave nonharmonically already at relatively low temperatures, which can only be treated reliably within the frame of molecular dynamics. Here the classical equations of motion are solved numerically for the N atoms and we get the positions, the velocities, and the accelerations for each atom of the system as a function of time.

As input we need the positions and the velocities of all atoms. The positions can be chosen in accordance with the ideal bulk structure, corresponding to a system consisting of a very large (almost infinite) number of atoms (see Fig. 4). For the velocities it is convenient to choose at the beginning of the simulation the same magnitude for

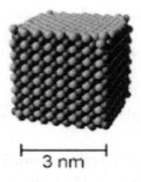

3 nm

Fig. 4   An aluminum nanocluster consisting of 500 identical atoms. It is the initial structural configuration, which is used as input for the molecular dynamics calculation. (© 2006, American Scientific Publishers.)

the velocities for all N atoms. However, the system develops toward equilibrium, i.e. toward Maxwell's distribution, after a relatively short time interval. More details concerning the distribution of the velocities are given in App. A.

The most important point is the interaction between the atoms. In the case of our Alcluster a realistic pair potential has been developed on the basis of the electronic structure of the atoms and the many-particle system, respectively. Here we do not want to discuss the details, but in App. B some general remarks in connection with interaction types are given.

Question: What happens with the Al cluster in the course of time during the molecular dynamics calculation? The result is surprising and remarkable; it is represented in Fig. 5.

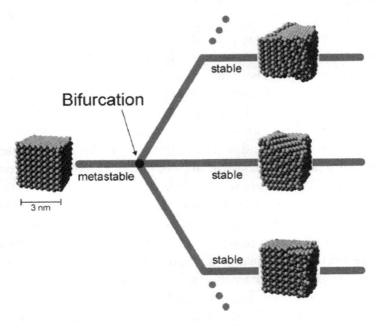

Fig. 5 There is a metastable clusterstate that transforms into a stable state, and there are various possibilities for that. It can in principle not be predicted into what stable state the metastable cluster transforms. At the bifurcation point ("Bifurcation"), nature plays dice in order to decide on which of the various branches the nanocluster will finally rest. (© 2006, American Scientific Publishers.)

2.1.1 Bifurcation Phenomena

Discussion of Fig. 5

Within the first time range, up to "Bifurcation," the cluster is in thermal equilibrium. Up to this point, the outer shape, the form of the cluster remains constant and so does the inner structure of the system. At the point "Bifurcation" the system, i.e. the nanocluster, transforms spontaneously without external influence; note that the cluster does not interact with its environment.

The cluster is obviously in a metastable state up to the point "Bifurcation" and transforms into a stable state. The following effect is of particular interest: when a cluster transforms from the metastable to the stable state, there are various possibilities for that. In other words, there is a *bifurcation* in the sense of chaos theory. At the bifurcation point ("Bifurcation"), nature plays dice in order to decide on which of the various branches the cluster will finally rest. No doubt, this is an interesting phenomenon.

The stable cluster configurations differ in both the inner structure and the outer shape. There are grain boundaries, dislocations, and other lattice defects. The outer shape of a cluster strongly depends on the arrangement of these lattice defects.

This bifurcation is an interesting phenomenon and there is no counterpart in micro- and macrotechniques. In macrotechniques, a system (for example a car) will always be destroyed after a certain time, and this degradation is due to the system's interaction with the surroundings, and has nothing to do with a bifurcation phenomenon. There are no *bifurcation phenomena* in micro- and macrotechnology.

As we have seen in connection with Fig. 5, the situation is different for nanosystems: the transition at the bifurcation point is constructive and new forms emerge, namely one of the possible stable states (two of them are represented once more in Fig. 6).

As said, in micro- and macrotechniques there are no such metastable states. If there were nevertheless bifurcation phenomena in the macroscopic realm, a screwdriver could transform spontaneously into a hammer or nippers, or a spoon could transform into a fork,

Fig. 6   At the bifurcation point new forms emerge that are stable (see Fig. 5). Two of them are represented here. (© 2006, American Scientific Publishers.)

and such transitions would take place without any external influence. This is of course nonsense!

Another example: we could put a penny on the table and this penny would transform spontaneously into a 50-cent piece. This would solve the current problems we have in the world, but such transformations are of course not possible.

All these phenomena, these *spontaneous transformations*, are not possible in micro- and macrotechniques, but should be real effects in nanotechnology, particularly within the frame of materials research at the nanolevel.

2.1.2 Analogies with Biological Systems

Although the Al clusters represent *inorganic* systems, there are certain analogies with biological systems. In this connection four points are relevant:

(1) A nanosystem can be creative; it can meet individual decisions that are not influenced from outside; the bifurcation phenomena can be interpreted in this way.

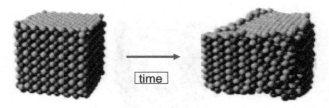

Fig. 7   At the bifurcation point, new forms emerge that are stable (see Fig. 5). Two of them are represented here. (© 2006, American Scientific Publishers.)

(2) A nanosystem is able to transform spontaneously; it can transform from a metastable state to a stable state without external influence. Spontaneous transformations are also typical phenomena within biology, particularly in connection with embryology.

(3) New forms, new shapes, which were not existent before, emerge without external influence. Also, within biology (for example in embryology) new forms appear, which were not existent before. This behavior reflects creativity.

(4) The shapes of such nanosystems transform to more complex forms (the stable state is more complex than the metastable state). In the stable case (Fig. 7, image on the right side), we need more description elements than in the metastable case (Fig. 7, image on the left side). This behavior is also typical of biological systems and is a typical feature of evolution.

Well-known biologists (embryologists) classified this Al cluster as a biological system. However, one should hesitate to draw such a conclusion. However, there is obviously no definite line between materials science and biology — at the nanolevel, of course. This point is of particular relevance and indicates that already relatively simple systems exhibit complex behavior with respect to the structure and dynamics.

Does there exist creativity at the nanolevel?

The bifurcation phenomenon reflects the behavior of single clusters, and this specific effect comes into play by the mutual influence of temperature and the interaction between the atoms forming the cluster. For the interaction, see in particular App. B. There is a certain kind of independent creativity, and this creativity does not come from outside, i.e. it is not initiated by the surroundings of the cluster or by the observer, but is an inherent characteristic of the system (nanocluster) itself.

Such a behavior is also known with respect to biological systems and supports the statements we have given above (see points 1–4). However, the usual biological systems are much more complex than the Al clusters studied here. Since this specific behavior (creativity) is essentially influenced by the temperature and the interaction potential between the atoms, nanosystems (nanoclusters) at zero temperature have to be considered as "dead" systems without any creativity. In conclusion, the question "Does there exist creativity at the nanolevel?" can be answered positively.

2.1.3 What is the Reason for Nanoeffects? A Few Remarks

Why do small systems (nanosystems) with a spatial extension of only a few nanometers behave differently from those we normally use in micro- and macrotechnology? The answer is relatively simple.

We have to distinguish between particles (atoms, molecules) at the surface of the system under investigation and those that are positioned in the bulk (the inner part) of the system. If the system has N particles and if N' of them are positioned at the surface or close to it, the ratio N'/N is the crucial parameter, which dictates the properties, i.e. the structure and dynamics of the entire ensemble. Let us briefly explain why.

The behavior of surface particles

The structure and dynamics of the atoms (molecules) in the surface region can be very different from that in the inner part of the system. The particles positioned in the surface region are in general less bonded than bulk particles.

The particles at the surface experience an asymmetric interaction situation. For particles with $r_L < r_c$ (r_L is the distance of the particle from the boundary of the array, and r_c is the range of interaction

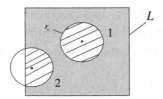

Fig. 8   Two of the N particles (called 1 and 2) of the system. The spheres around each particle are the interaction ranges, having in both cases a radius of r_c. The interaction is effective within the hatched areas of the spheres. The influence of the interaction is asymmetric (atom 2), and it is symmetric in the inner part of the system (atom 1).

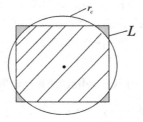

Fig. 9   The same situation as in Fig. 8. However, here the extension L of the system is smaller than in Fig. 8 but it is assumed that there is the same cutoff radius r_c for the interaction potential. All N atoms behave like surface atoms, i.e. none of them can be in the bulk state.

around particle $i, i = 1, \ldots, N$), the interaction with their surroundings is asymmetric (see Fig. 8, particle 2); on the other hand, it is symmetric for particles with $r_L > r_c$ (see Fig. 8, particle 1). This asymmetric situation has the effect that the particles near the surface are less bonded than those in the bulk of the material. If we have $r_L < r_c$ for all particles, none of them experience a symmetric interaction range (except for the atom that is exactly positioned in the middle of the system). The situation for $r_L < r_c$ is represented in Fig. 9.

Again, the surface particles are in general less bonded than bulk particles. This fact leads to anharmonic effects at the surface, and the mean square displacements become relatively large here; the vibrational amplitudes at the surface are distinctly larger than in the inner part of the system.

Description of anharmonicities

What does anharmonicity mean with respect to the forces acting between the particles? The atom vibrates around a certain position. This is the so-called equilibrium position. We talk from anharmonic effects, when the restoring force F, acting on the atom, does vary *linearly* with the displacement $r = (x, y, z)$ from the equilibrium position. In other words, when F is proportional to the displacement x (we consider only the x-component), $F \propto x$. In this case the description of the dynamics of the atoms (molecules) becomes relatively easy and can be treated within the usual solid state physics.

In the case of anharmonicities, we have higher-order terms $x^2, \ldots$ and we get $F \propto x, x^2, \ldots$ instead of the law $F \propto x$. In the usual solid state physics, we work always within the harmonic approximation; there are only a few exceptions. Just the opposite is the case in nanoscience and nanotechnology. Here the typical effects can only be understood by consideration of the anharmonic terms, even at low temperatures.

Effects due to the anharmonicities

These anharmonic effects can cause the particles at the surface to perform a diffusive motion, i.e. these particles are molten, which is known as the effect of premelting. In other words, the crystal is in the surface region in the liquid state already below the melting temperature.

Since the relative number N'/N of surface atoms (molecules) increases with decreasing size L of the system, these effects (premelting, etc.) are often very large in the case of nanosystems and the melting temperature of the whole nanosystem is in general far below the melting temperature of the material in the bulk. This is actually the case for the Al cluster we have studied in this chapter (see Fig. 4), which consists of only 500 atoms. In this case N'/N is not far from the number 1, i.e. almost all Al atoms behave like surface particles.

This in particular means that the basic principles of solid state physics can hardly be used or extended to the realm of systems that are of nanometer size. For example, the concept of phonons cannot be extended to nanosystems, and we have to use other methods for the determination of the particle dynamics of such systems. Only a few methods, which work well in the description of bulk states, can be used for the characterization of nanosystems. In conclusion, the typical features of the solid in the bulk (periodic structure and harmonic behavior) can be no basis for the theoretical description of nanosystems, and the system under investigation has to be studied carefully with respect to an adequate theoretical treatment.

The situation is entirely different in micro- and macrotechnology. Here we have approximately 10^{10}–10^{22} particles, and N'/N becomes much smaller than 1: $N'/N \ll 1$. In these cases the effects due to the surface atoms can be neglected; the number N' of surface atoms is much smaller than the total number N of atoms.

2.1.4 Interaction Potential for Al

In the last chapter we have learned that the temperature dependence of system properties at the nanoscale (for example the properties of Al clusters which we have discussed above) is an essential factor and has to be considered carefully. What about the interaction potentials in metallic nanosystems? Have they to be treated as temperature-dependent quantities? The answer is yes.

In the case of noble gases, the temperature dependence of cluster properties[4] can be treated in a good approximation on the basis of *temperature-independent* pair potentials. The reason is that van der Waals interactions (see in particular App. B) are not sensitive to variations in temperature. This is in general not the case for metallic systems, for example Al. Why?

A metal consists of ions and conduction electrons (see also App. B). We always considered above the structure and dynamics of an ensemble (cluster) of Al ions. The interaction between the ions is strongly influenced by the *screening* of the conduction electrons. The screening is sensitive to the density of the conduction electrons within the system under investigation, and this electron density is temperature-dependent. Therefore, we may conclude that the pair potential in metals is in general dependent on temperature.

How relevant are these effects? Let us give some details, but only superficially.

We can state quite generally that the electron properties within many-body systems and nanosystems are reflected by the famous dielectric function $\varepsilon(q)$, which is also an essential quantity for the determination of the pair potential within the framework of pseudopotential theory.

How large are such temperature effects for Al? In a first attempt[5] we used a pseudopotential by Dagens *et al.*[6] This potential is a nonlocal pseudopotential; the screening is described by the Geldart–Taylor dielectric function. However, this potential was not sufficient for describing the structure and dynamics of the Al atoms (Al ions) at the surface. Therefore, we used another model for the pseudopotential which is obviously more realistic.[7] In this model, the essential point is the consideration of interactions of the van der Waals type (see App. B) in the determination of the direct ion–ion potential as well as in the ion–electron part of the interaction. (Numerical calculations for liquid rubidium near the melting point showed that van der Waals-type interactions are distinctly reflected in the metal potential.[8])

The pair potential for Al based on this kind of potential is shown in Fig. 10 for two temperatures (300 K and 1000 K; the melting temperature of Al is 933 K). The free parameters have been fitted with

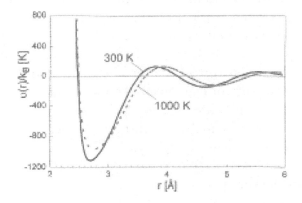

Fig. 10 Pair interaction potential for Al at 300 K and 1000 K; for comparison, the melting temperature is 933 K Although the temperature effects are relatively small, they have to be considered in an adequate description of Al-data.(© 2006, American Scientific Publishers.)

the help of crystal and liquid data for various temperatures and, therefore, the temperature-dependent interaction potential in the figure describes a wide range of experimental data and surface properties:

(1) The crystal is stable up to the melting point. The phonon density of states at $T = 300$ K is described well.

(2) The structure and dynamics of the liquid state are also described adequately by the potential.

(3) The melting temperature (933 K) of Al is well represented by the potential.

(4) We found for the outermost layer an inward relaxation, which is confirmed by experiment.

(5) The mean square displacements at the surface agree well with LEED data.

(6) The molecular dynamics results indicate that there are relatively strong anharmonic effects already at low temperatures; this is confirmed by experiments too.

(7) The onset of premelting ($T > 650$ K) which results from the molecular dynamics calculations agrees excellently with the experimental observations.

As can be seen from Fig. 10, the temperature effects in the potential are relatively small. However, they have to be considered in an adequate description of Al data. This indicates that the properties of many-particle systems (particularly nanosystems) are very sensitive to small variations in the interaction potential. This is the reason why the potentials have to be determined very accurately and carefully, and this has been done for Al; the deduced potential (Fig. 10) correlates a large number of experimental data, which are listed above (points 1–7). The procedure used for Al (construction of a temperature-dependent pair potential and realistic adjustment to experimental data) can be considered as exemplary.

2.1.5 Temperature Effects

As we have discussed above, the temperature of nanosystems is a crucial parameter; the properties of nanosystems vary strongly with temperature — much more than in the bulk of the crystal. This fact can have the effect that certain nanosystems (nanomachines, etc.) work properly only within a relatively small temperature window, because components (or even the whole nanosystem) lose their stability outside this temperature window.

The reason for this temperature sensitivity is obvious and can be understood from the fact mentioned above: since the atoms at the surface are in most cases less bonded than the atoms in the inner part of the system (when they are in the bulk state), anharmonicities come into play, whereby the intensity of these anharmonicities varies strongly with temperature. Therefore, we have to expect that the structure and dynamics of the surface particles are more sensitive to temperature variations than the corresponding properties in the bulk. Since we have a lot of surface states in connection with nanosystems (see in particular Figs. 8 and 9 and the corresponding text), the properties of the whole nanosystem have to be considered as being strongly dependent on temperature. This is, for example, the case for the mean square displacements. However, in addition, the effect of premelting, which is very often observed even for macroscopic systems, is driven by strong anharmonicities at the surface. Here premelting again means that the macroscopic crystal is molten in the surface region already below the melting temperature.

Because the relative number of surface atoms increases with decreasing size of the system, these effects are large in the nanometer realm and, as has already been remarked above, the melting temperature of the whole nanosystem can be distinctly below the melting temperature of the same system in the bulk.

The temperature behavior of nanosystems is of particular relevance. The interplay of temperature and the interaction between the atoms forming the cluster (their mutual influence) can lead to a certain kind of creativity in analogy with what is observed at the biological level. This has been demonstrated in connection with the bifurcation phenomenon with respect to Al clusters.

Also, this behavior of nanosystems makes clear that the basic principles of solid state physics can hardly be extended to the realm of systems that are of nanometer size. For example, the concept of phonons cannot be extended to nanosystems, and we have to use other methods for the determination of the particle dynamics in such systems. Due to the anharmonicties, the concept of phonons has to be generalized adequately.

We may state quite generally that only a few concepts, which work well in the description of bulk states, can be used for the characterization of nanosystems.

2.1.6 Methods of Description

Properties as a function of the particle number

In conclusion, the typical features of the solid in the bulk (periodic structure and harmonic behavior) can be no basis for the theoretical description of nanosystems. The system under investigation has to be analyzed carefully with respect to the theoretical treatment. The principles of modern surface physics are often useful within the analysis of nanometer-scale properties. The reason is obvious. In the case of nanosystems, a great fraction of atoms belongs to the surface region of the system and determines essentially their behavior, and such nanosurface effects can be much larger than those of semi-infinite surface systems. For example, in the case of sufficiently small nanoclusters no particle of the system can be considered as a bulk particle since the distances of all the particles to the surface are smaller then the cutoff radius of the interaction (see also Fig. 9).

Nanosystems can have properties which are distinctly different from those of macroscopic systems: properties that are well defined and clearly fixed in the bulk are very often no longer typical features in nanophysics and nanotechnology. If the particle number is decreased so that the size of the system is in the nanometer region, new effects emerge which have no counterpart in other fields of physics, chemistry, and technology.

We have already discussed above that the melting temperature of a macroscopic system is usually well defined, and the system is thermally stable up to the melting point. Of course, there can be phase transitions, but the various phases define stable configurations. In contrast to macroscopic systems, the melting temperature of a nanosystem in general depends on the particle number and is also a function of the outer shape of the system. Moreover, the melting temperature of certain nanosystems is not defined and not clearly fixed, which can be explained by the faster occurrence of sublimation compared to the melting process.

New situation

This is a completely new situation, particularly in connection with materials research. Clearly, the construction of nanomachines and other aggregates is based on components whose material properties have an essential influence on the function of such systems. Therefore, the definition and classification of material properties at the nanolevel is a particular challenge and is just at the beginning. Within such a program not only further sophisticated experiments have to be conducted, but also specific methods for the theoretical treatment of such systems have to be selected and refined carefully.

Nanotechnology will also lead to drastic new insights in connection with biological systems, not only by the manipulation of the features of existing biological structures but also with respect to questions such as "What is a biological system?" or the quite general "What is life?" It is usually assumed that biological features are strongly correlated with systems of sufficiently large complexity, particularly, in connection with complex molecules. But do we really need complex systems and complex organic molecules, respectively, for all phenomena which arise in connection with biological (living) systems? Obviously, this must not be the case: as we have recognized with regard to the Al cluster (Fig. 4), a sufficiently small system made up of an inorganic monatomic material (aluminum) shows surprisingly new effects in analogy with those which are typical of biological systems. How far has such an Al system to be reduced to show such features? In the micrometer realm the system behaves completely non-biologically. However, if the size is of the order of a few nanometers, such new effects emerge.

2.1.7 Concluding Remark and Summary

Nanosystems are interesting and behave in relevant cases quite differently from systems used in micro- and macrotechnology. However, many researchers discuss them on the basis of traditional thinking. Why? They very often study the effects by means of static building blocks, as we do in connection with micro- and macrosystems. Normally it is thoroughly assumed that the properties of nanosystems (for example a nanocluster) are due to the relatively large surface area; the relative number of surface atoms increases with decreasing size. This effect modifies the structure of the building blocks (in comparison with that in the bulk), but they remain static elements within this view.

However, this structure aspect is only one typical effect in connection with nanosystems. There is another important point: nanosystems behave strongly anharmonically, i.e. the vibrational amplitudes are relatively large and the harmonic approximation is hardly applicable here. These anharmonicities are in the case of nanosystems (particularly with regard to nanoclusters) much more pronounced than in the case of micro- and macrosystems. (We have already discussed this point above.) In other words, the dynamics is also modified, which can have the effect that the whole nanosystem can transform spontaneously and there can be many co existing structural states (see Fig. 5). We have briefly discussed a typical nanocluster and we considered it as a relevant example.

Nanoclusters can be in an excited state (like atoms). After a certain time the excited cluster transforms spontaneously to the ground state without external influence. There can be more than one ground state and it is quite a matter of chance to what ground state the cluster transforms from the excited state. The various ground states may differ in both the inner structure and the outer shape. There are grain boundaries, dislocations, and other lattice defects. The outer shape of the clusters in the ground state depends on the arrangement of these inner lattice effects.

In other words, a free cluster or a cluster on a surface is for a certain time interval a static system. But suddenly it transforms spontaneously into another system, with another inner structure and another outer shape. In micro- and macrotechniques we do not observe such effects; here we have for all times one structure and one shape. A penny on a table does not transform spontaneously into a 50-cent coin; a penny remains a penny forever. However, such penny (nanopenny) transformations would be possible on nanometer length scales.

We may conclude that the behavior of systems — relevant in nanotechnology — is in general complex and is obviously not comparable with the behaviour in micro- and macrotechnology. For the description of such nanosystems the theoretical and computational tools have to be selected very carefully. In this connection the interaction laws (potentials) between the atoms forming a nanosystem are critical functions because the structure and dynamics of such systems are very sensitive to small variations in the potentials. We have discussed this point above.

2.1.8 What is Important?

In the description of nanosystems three points are of considerable importance: the particle number N, the interaction potential between the particles, and the temperature of the system under investigation. Let us summarize the main facts:

(1) The number N of particles is relevant, and this is because with decreasing number N the relative number of surface particles increases. The surface particles behave quite differently from those in the inner part of the system.

(2) The interaction potential between the atoms (ions, molecules) plays a basic role in the description of nanosystems, and it has to be determined very carefully. The potential is in general temperature-dependent, as we have outlined in connection with Fig. 10. The properties of nanosystems are very sensitive to small variations in the interaction potential. These effects can be drastic and have to be considered for each case (the system under investigation) separately and carefully.

(3) The temperature of nanosystems is a crucial parameter, because in nanoscience the properties of systems vary strongly with temperature, much more than in micro- and macrotechnology.

No doubt, these three points are relevant and, furthermore, we always have to keep in mind that nanosystems behave in most cases strongly anharmonically. Such anharmonicities are important and have to be treated carefully within the frame of theoretical investigations. In particular, it has turned out that the anharmonicities cannot be treated as small perturbations to the harmonic approximation. Also, this fact is often underestimated.

2.1.9   Solving the Equation of Motion

How can we determine the properties of a system consisting of N particles? Here we use and solve, in the classical case, Newton's equations of motion; the atoms of most systems behave classically for relevant temperatures.

In the conventional treatment of this problem, i.e. in the solution of the equation of motion, we need "simple models." We may state quite generally that simple models can be obtained by controlled approximation from the general formulation of the equations of motion. With regard to many-particle systems, the general mathematical formulation of the problem is in most cases so complex that simplifying models have to be chosen, but very often they cannot be obtained by controlled simplifying steps from the general cases. It is therefore a rule to introduce simple models just for convenience (see also App. C).

On the application of simplifying models

The "simple model" of solid state physics is the crystalline solid in the harmonic approximation. On the basis of this model, one is able to determine successfully the properties of a lot of materials. However, there are also a lot of cases where this simple model is not applicable, even when it is extended by specific assumptions. For example, the silver subsystem of the solid electrolyte α-AgI is highly disordered and shows strongly anharmonic behavior; a simple model for α-AgI and similar materials is not available.

For liquids and gases, simple models have not been found either. Even in the case of gases with low densities, we cannot simply restrict ourselves to the first terms in the virial expansion in the calculation of the pressure. This is because the expansion obviously converges slowly and, therefore, even in the case of low density, one has to consider more than the first two terms. The virial coefficients can be expressed in terms of the pair potential $v(r)$, but such expressions are getting complicated for the higher-order virial coefficients and in practical calculations only the first terms are accessible. The virial expansion would define a simple model if for a broad class of gases a restriction on the first two terms would be realistic.

In summary, only for a specific class of many-particle systems could a simple model be found: it is the crystalline solid in the harmonic approximation, but in most cases anharmonicities cannot be considered as small perturbations to the harmonic approximation. This is also the case for typical nanosystems, since a great fraction of atoms is more or less close to the surface region of the system, and surface particles behave strongly anharmonically. In small nanosystems (for example clusters with a few hundred atoms; see Figs. 4 and 9), even the innermost particles cannot be treated as bulk particles since the cutoff radius of the interaction potential is larger than the distance to the surface. The atoms at surfaces are less bonded than in the bulk and, therefore, the mean square amplitudes of the atoms are significantly larger than at the surface, leading to relatively strong anharmonicities. This phenomenon can be observed even at low temperatures — low in comparison with the melting temperature of the bulk system.

Basic information

Only on the basis of the general formulation can the relatively strong anharmonicities be treated without approximation. In other words, the general description of the properties of classical many-particle systems (for example nanosystems) can be given by means

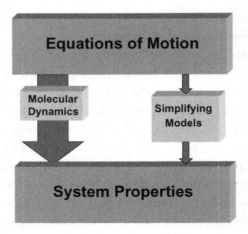

Fig. 11   If a "simple model" is available, the properties of many-particle system — including those of nanometer size — can be determined on the basis of such a simplifying model; in most cases, additional assumptions are necessary. The advantage of molecular dynamics is that the description can be done without simple models and other additional assumptions. This is important since in most cases such systems are theoretically so complex that simplifying models have often been chosen for convenience without the use of controlled simplifying steps.

of molecular dynamics calculations without the use of "simplifying models" (Fig. 11) or other simplifying assumptions.

As mentioned above, such simple models and simplifying assumptions, respectively, are not known at the microscopic level and can only be introduced in a phenomenological or empirical way. As remarked above, such simplifying models often had to be chosen for convenience and could not always be obtained by controlled simplifying steps from the general equations of motion. Within molecular dynamics we are able to work without the use of simplifying models.

The molecular dynamics method is a powerful numerical method, and it is relatively easy to understand the procedure. Within molecular dynamics the classical equations of motion are solved by iteration with the help of a high-speed computer. In this way we obtain the coordinates and the velocities (momenta) as a function of time for all N atoms (ions and molecules). This is the complete, basic information about the structure and dynamics of the N-particle system. More details are given in App. A.

2.1.10 On the Design of Specific Nanosystems

Ferroelectric nanodevices

The development of novel nanodevices using the functionality of ferroelectrics is essential and interesting. Pervoskites exhibit ferroelectric behavior with the coexistence of several effects that are dependent on the temperature, applied field, and strain.

This is the basis for the development of multifunctional ferroelectric nanodevices. The theoretical investigation of such systems requires the application of a great variety of methods in order to be able to investigate the electronic and structural properties of such nanoscale ferroelectrics.

Here the molecular dynamics and the Monte Carlo method have to be applied, but also the density functional theory (DFT) for the study of ground state properties and the time-dependent DFT for the investigation of excited state properties. In addition, growth models have to be developed.

Design of specific materials

The mobile device industry is interested in materials with specific properties. What kind of materials? There is a need for materials with tuneable dielectric permittivity (with tuneable magnetic permeability) and for materials with small losses in the GHz range. In this connection magnetic nanometer-scale particles are of interest — to coat them with a suitable dielectric, and to embed these coated particles in a medium with minimal losses in the GHz range.

Such a technological program can only be understood in terms of basic physics: nanoparticle assemblies have to be modeled and we need here potentials for the description of the particle–particle interaction and the particle–matrix interaction. The coating of the particles has to be described by microscopic electronic structure calculations and molecular dynamics.

In the modeling of transport phenomena we have to include the relevant quantum aspects at the atomic scale and to extend them by multiscale modeling to macroscopic leads and contacts; here also finite element methods come into play.

Low density lipoprotein

The application of the theoretical physical methods is also of particular relevance to medicine. Let us briefly discuss an example. The main cause of cardiovascular diseases is atherosclerosis, a disease in which so-called LDL particles accumulate in the innerpart of the arterial wall. LDL means low density lipoprotein, in which at least one of the components is a lipid. Lipoproteins are the typical means by which lipids are transported in the blood.

In this connection the functional properties of certain nano domains on the LDL particles are of particular interest. The accumulation of LDL particles in the innerpart of the wall has been studied at the basic level, i.e. by molecular dynamics. As input information we need here the interaction potentials between the LDL particles as well as the interaction between the LDL particles and the wall.

Biomolecular nanomotors

In an adequate description of biological molecular motors or nanopores, the techniques of theoretical physics are also of principal importance. A biological molecular motor can be a protein that converts chemical energy into mechanical work and could be used as an actuator; such motors are smaller than 100 nm. Nanopores are responsible for selective transport of molecules or ions between different compartments; only a little is known about the transport of macromolecules through nanopores.

Also, the combination of molecular motors with other nanodevices (for example nanopores) would be of particular relevance to basic research as well as to applications. The theoretical and computational analysis of such systems has to be done at the highest level, i.e. we have to apply molecular dynamics, molecular mechanics, Brownian dynamics, Monte Carlo methods, DFT calculations, and of course multiscale modeling.

Avidin–biotin technology

Certain proteins can be employed for the construction (design and production) of new and useful materials (nanomaterials) — for example, the use of avidin. Avidin is a chicken protein having a diameter of approximately 5 nm. The polymer avidin consists of four identical monomers and each of them can bind one biotin molecule. Avidin has an important property: it has an extreme affinity with biotin (D-biotin) and other ligands.

The use of avidin molecules, together with certain ligand combinations, can lead to a completely new level of functionality, and this can open the door to totally new applications. Avidin–biotin complexes have been used in numerous applications within the so-called avidin–biotin technology.

For example, this technology has been employed to modify carbon nanotubes to form transistors by self-assembly. Furthermore, certain building blocks can be used for the construction of two- and three-dimensional complexes, also by self-assembly. All these points could be of interest for the development of future nanodevices and nanomaterials.

Biomolecular translocations

An important topic is the theoretical and computational investigation of the biopolymer translocation through nanopores. The study of such translocation processes is of basic interest and important in connection with biotechnological applications, such as, molecular filtering or protein transport through membranes, and also with respect to injection of viral DNA into host cells.

It is essential that such a translocation process is a multiscale problem. On the one hand, the understanding of the microscopic details (structure and dynamics) of the polymer–pore system is necessary. On the other hand, processes on macroscopic timescales also have to be understood (for example the translocation time). One can do it in a bottom-up approach; one starts with microscopic models (usually on the basis of molecular dynamics calculations). This microscopic information will be the basis for coarse-grain models for such polymer–pore systems which have to be constructed carefully.

Hierarchical nanostructures

So-called hierarchical nanostructures can show unique properties. The next step after self-assembly is hierarchy, where the primary self-assembled building blocks associate into more complex secondary structures that are integrated into the next size level in hierarchy. Hierarchy is a characteristic of many self-assembling biological structures and shows relevant properties that are not found by self-assembling alone.

As a possible application the optical and photoactive properties of artificial hierarchical nanostructures could be of interest. Such knowledge would open the door to nanoscale engineering of materials with predefined optical and photoactive properties for applications in solar cell technology and nonlinear optics.

In the theoretical and computational analysis of such processes, we need techniques that cover a wide range of length and time scales. Here, too, multiscale modeling (in connection with atomistic simulations and coarse-graining approaches) is relevant, but also density functional studies for the determination of ground state properties. Time-dependent density functional studies are necessary for the determination of excited states for the description of radiation effects.

Nanoparticles as vectors in gene delivery

The use of nanoparticles as vectors in gene delivery is a particular challenge. In conventional gene delivery, viral vectors are employed. But there are some disadvantages in connection with this conventional method. For example, viral vectors are immunogenic. However, nanotechnology opens the door to tailoring therapeutic gene delivery systems where nanoparticles function as vectors. Also here the theoretical tools for the description of such systems have to be chosen adequately. Such models are in general complex: for example, the core of the nanoparticle can be made up of solid DNA. These cores can be coated with specific polymers. Then, water diffusion into the core (DNA system) leads to nanoparticle bursting; bursting is for the release of DNA in the host cell. This process of bursting is relatively complex and has to be modeled.

Mechanical properties of nanostructures

Quantum dots are relevant functional elements in connection with optoelectronics. Such optoelectronic properties can in principle be modified by imposing various stress conditions on the system using nanoidentation. It is a challenge to describe such phenomena theoretically (computationally). Stress-affected effects are of particular interest for low-dimensional materials, i.e. for zero-, one-, and two-dimensional systems.

Quantum dots are zero-dimensional systems but also nanoballs with a diameter range from 5 to 100 nm. Examples of one-dimensional nanosystems are nanotubes, nanorods, and nanobelts. Thin films and multiple-layer systems are examples of two-dimensional systems.

Conclusion

All these examples demonstrate that we have to work at the highest theoretical level in nanoscience and nanotechnology, and this level is defined by the laws of theoretical physics.

2.1.11 Theoretical and Computational Methods

General remarks

The systems we have discussed in Sec. 2.1.10 are relatively small, but complex. Simple analytical models for certain classes of nanosystems could not be found so far. Specific simulations techniques are therefore mostly used in the understanding and prognosis of such nanosystems. However, the simulation models have to be prepared very carefully in order to be sure that all relevant features and mechanisms of the complex nanosystem under investigation are considered.

In other words, with regard to simulations the input is important and has to be investigated critically. Here it is essential to choose adequate physical equations but also realistic physical boundary conditions. In many cases a lot of simulation techniques and theoretical methods are mentioned. But almost nothing is said about the nanospecific aspect.

Just at the nanolevel, the usual theoretical (computational) methods have to be developed further and it is of particular importance to recognize the weak points of these usual methods when they are applied to nanotechnology.

In most cases these methods have been taken over from other disciplines (solid state physics and surface science) without considering the nanospecific features. Question: Can all theoretical and computational methods be used without nanospecific modifications? There is no rule for answering this question generally, but each specific nanoproblem has to be analyzed with respect to this point. This has to be done carefully, even when the methods work excellently in the bulk or in the case of micro- and macrosystems. In the following, the most relevant methods for the theoretical treatment of nanosystems are mentioned.

The most important techniques

Let us mention here the most relevant theoretical and computational methods that are important for the description of nanosystems. As already said, analytical models for the complete description of the structure and dynamics of nanosystems are not known since there are obviously no specific concepts or simple models for that. In solid state physics the crystalline solid in the harmonic approximation can be considered as such a specific concept. However, the harmonic approximation has almost no basis in the theoretical and computational treatment of typical nanosystems and, therefore, anharmonic effects can only be treated numerically. Thus, adequate description of systems with strong anharmonicities requires numerical investigations, i.e. we must have recourse to simulation techniques. Here the molecular dynamics method is of particular importance. But there are still other numerical techniques.

Let us briefly discuss the most relevant tools — molecular dynamics, quantum molecular dynamics, nonequilibrium molecular dynamics, and the Monte Carlo method — and we will also make some brief remarks concerning multiscale modeling. Other important methods used in nanoscience are finite element methods, Brownian dynamics, Langevin dynamics, molecular mechanics, etc.

In connection with electronic properties the following approaches are relevant: density functional theory, time-dependent functional theory, Hartree–Fock approximation, and the potential morphing method, which can be considered the most powerful method for solving Schrödinger's equation.

2.1.12 Features of the Methods

Molecular dynamics

This method and its relevance have already been discussed above, particularly in Sec. 2.1.4. The relatively strong anharmonicities in nanosystems are not negligible even at low temperatures, and we have to describe such systems on the basis of the most general formulation, namely by the numerical solution of the equations of motion. More details on molecular dynamics are given in App. A.

Nonequilibrium molecular dynamics

A few remarks should be made in connection with nonequilibrium molecular dynamics, which has been developed in addition to the usual (equilibrium) molecular dynamics and has been known since the early 1970s. This method was introduced for efficient computation of transport coefficients. To establish the nonequilibrium situation of interest, an external force is applied to the system. Then the response of the system to these forces is determined from the simulation. This method has been used for the calculation of diffusion coefficients, shear and bulk viscosity, and thermal conductivity.

Non-Hamiltonian systems can be studied by means of nonequilibrium molecular dynamics, for example dissipative systems, i.e. units that involve friction in one of its various forms. In such cases particle trajectories are calculated from the equations of motion, which are, however, not consistent with any Hamilton function.

Quantum molecular dynamics

A method that directly combines quantum theory with classical mechanics has been developed by Car and Parinello. The equations of the density functional formalism[16,17] are solved simultaneously with the classical equations of motion. Within this method (quantum molecular dynamics) the classical Lagrange equation of the atomic positions and velocities is extended by a fictitious dynamics of the Kohn–Sham wave functions and their time derivations. Calculations of this type are relatively complex and, therefore, the quantum molecular dynamics method is restricted to a few hundred particles. The method can lead to relevant statements but, unfortunately, it has not been used very often.

The Monte Carlo method

In contrast to molecular dynamics, where we have completely deterministic algebraic equations, the Monte Carlo method is a numerical approach in which specific stochastic elements are used. Whereas the molecular dynamics method enables the investigation of static and dynamic properties, the Monte Carlo method only deals with static properties; these are configurational averages and thermodynamic quantities.

The results produced by the two methods (molecular dynamics and the Monte Carlo method) should be in agreement to the order of N^{-1} and, therefore, one can expect agreement of the two methods within statistical error.

The Monte Carlo method has been employed in a diverse number of ways. The following types have often been used: classical Monte Carlo, quantum Monte Carlo, path integral Monte Carlo, volumetric

Monte Carlo, simulation Monte Carlo, inverse Monte Carlo. Here we will not give details about this diverse number of ways, but each type of method has its justification.

Multiscale modeling

Multiscale modeling helps us to understand how nanoscale properties give rise to large-scale behavior. Certain strategies have to be chosen in order to be able to study the properties over a wide range of length and time scales, and this requires multiscale modeling. Here atomistic and coarse-grain simulations are at the center and have to be done for multiscale modeling.

In materials science, multiscale modeling is used to predict and explain the properties (for example the mechanical properties) of a certain material at dimensions ranging from a fraction of a nanometer to a meter. Here three length scales are of particular interest: atomic scale (nanometer), microscale (micrometer), and mesoscale (millimeter and above). Multiscale modeling makes connections among these scales. Such investigations are exclusively based on computer studies.

Multiscale modeling: a typical example

In connection with multiscale modeling, we have already discussed an interesting example: The biopolymer translocation through nanopores (Sec. 2.1.10). This translocation process is a typical multiscale problem. On the one hand, the understanding of the microscopic details (structure and dynamics) of the polymer–pore system is necessary. These processes (such as molecular filtering or protein transport through membranes) take place on a large timescale, i.e. not only is the microscopic behavior relevant but also processes on macroscopic timescales have to be understood (for example the translocation time). As has been mentioned, one can do it in a bottom-up approach starting with atomistic models (usually on the basis of molecular dynamics calculations). This microscopic information will be the basis for coarse-grain models for such polymer–pore systems. Clearly, there are other relevant applications, but we will not deepen this point here.

2.2 FOOD CHEMISTRY: WHAT IS A NANOPIZZA?

No doubt, the behavior of the Al cluster is astonishing. That such inorganic systems would behave like biological systems was far from our everyday life imagination. Let us leave the cluster story and come to another surprising example, from chemistry — or, more precisely, let us enter the realm of food chemistry. Here a lot of changes could be come up to us. As a typical example we mention what is called in the literature "nanopizza".

We all know what a pizza is. When we visit an Italian restaurant, the menu often offers more than 20 pizza types: a pizza made of cheese, a pizza made of spinach, a pizza made of fish, and so on and so forth.

This is common and nothing new. However, nanotechnology could change this situation fundamentally. In the future we will possibly have all these pizza types in one. What does this mean? How can we produce in the future all the pizza types on the basis of one?

In the not-too-distant future we will be able to buy a new type of pizza, namely a nanopizza, with the following property: If the pizza is warmed up, for example to 100°, we get a spinach pizza. If, however, the same pizza is warmed up to 150°, we receive a fish pizza. Within other temperature ranges, other pizza types are created. Unbelievable!

No doubt, this is vertiginous and makes the extent of changes clear, which come up to us. All that is far from our usual experience, but should be realizable in the not-too-distant future. Sure, it is a strange scenario, but also here we must say that everything is a matter of habituation. Nevertheless, we have to ask whether we really want such more or less artificial things from food chemistry.

2.3 MEDICINE; BRAIN RESEARCH

The phenomena in connection with the nanopizza, just discussed, are without any doubt astonishing. In the future, the new products of food chemistry, created at the nanolevel, will lead to similar strange things and meals, respectively.

Here we only talk in the first place from principal possibilities. Whether we really want such artificial nanoproducts remains at this stage an open question, but we can be sure that it will lead to controversial debates.

On the Impact of Nanotechnology

The impact of nanotechnology will be tremendous. The manipulation of atoms, molecules, etc. will allow us to construct new technological worlds and will bring fundamentally new possibilities in the field of medicine and, as we recognized above, with respect to food production.

Just in the case of nanobiotechnology, big changes are expected already in the near future. It has been speculated that through nanotechnology our bodies will be transformed into illness-free, undecaying systems of permanent health. Moreover, it has been prognosticated that it will take approximately 30–50 years to develop nanotechnological means for the creation of superhuman intelligence. That will bring the human era to an end: "We already have experimental hints of how that might occur. In September 1999, molecular biologist at Princeton reported adding a gene to a strain of mice, elevating their production of NR2B protein. The improved brains of these 'Doogie mice' used this extra NR2B to enhance brain receptors, helping the animals solve puzzles much faster. A kind of genetic turboaccelerator for mousy intelligence. Human brains, as it happens, use an almost identical protein. It is not far-fetched to suppose that we will learn to tweak or supplement it to increase our own effective intelligence (or that of our children)."[9]

The following question arises: Are the theoretical tools, developed so far, sufficient for understanding this kind of experimental manipulation? Are we really in a position to describe biological phenomena (see also Ref. 9)? We will discuss this point critically, by means of basic principles.

2.3.1 Brain Research

Without doubt, the nanoeffects in connection with food chemistry are more drastic when we enter the realm of brain research. The specific experiment for increasing mousy intelligence is very basic and it is obvious to transfer the basic idea to human beings. In fact, similar experiments are explicitly planned for man. Is that possible at all? What are the arguments?

In order to be able to answer these questions, let us summarize the main facts. There is a certain kind of genetic turboaccelerator for mousy intelligence. The molecular biologists at Princeton added a gene to the brain of a mouse, elevating the production of NR2B protein. The improved brains of the mice used this extra NR2B protein to enhance brain receptors, helping the animals to solve puzzles much faster. It is of particular interest that human brains use an almost identical protein. Therefore, it is obvious to assume that we will learn to increase our own effective intelligence or to cure Alzheimer's disease (in the case of another treatment, of another input).

Again, the following question is essential: Are the theoretical tools developed so far sufficient for the description of such biological phenomena? In the following we will give some basic arguments for the view that only restricted statements are possible when we try to analyze the brain functions of a human being.

Self-organizing process

We have to be careful when we try to change our brain. What happens? We add the gene to the brain and a self-organizing process starts until the brain is in the new, improved state.

But is it always an improved brain state? This is of course dependent on what we add to the brain. In principle, the self-organizing process could also lead to a disaster because, due to the process, the healthy brain functions could be disturbed simultaneously or could even be destroyed.

In order to avoid that, the self-organizing process has to be investigated *before* we start this kind of activity. However, what are the tools for such investigations? What in particular is the level of description? In the next section, "levels of reality" and Gödel's theorem are of principal relevance when we try to analyze the situation in connection with brain functions.

The following question arises: Do we really know what a brain is, and how it works? In other words, is a human able to understand his own brain? Can a human make statements about a brain that is more complex and more powerful than that given by his present brain structure? Here Gödel's theorem comes into play. This famous theorem should also be applicable to brain functions.

2.3.2 Levels of Reality

We probably have to leave the level of the present physical description, where the material objects and their interactions are described by the tools of classical mechanics and the usual quantum theory. In what direction do we have to go? There is a principal question that has to be answered: Can we really develop the theoretical means for the description of superhuman intelligence?

To do that experimentally, i.e. in a practical way, is probably no problem. But can we really understand such a step? In other words, can we fully estimate the consequences for our life and civilization? This seems to be more than doubtful. The present theoretical tools and conceptual strategies are probably not sufficient for that. It is questionable whether we can understand a superhuman intelligence on the basis of the present human brain structure. Let us briefly discuss why.

On the description of brain functions

To clarify the situation in more detail, we have to introduce a useful and convincing concept about what we call reality and its description on certain levels. The following view has often been discussed in the literature[10]: Basic reality, i.e. reality that exists independently of the observer, is principally not accessible in a direct way. But it is observable or describable by means of pictures (pictures of reality)

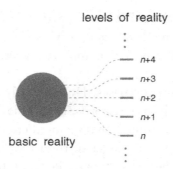

Fig. 12   In principle, we cannot make any statement about basic reality (actual reality outside), which is symbolized in the figure by the full circle. But we can observe or describe aspects of it within the framework of "levels of reality" which are vertically arranged in accordance with the degree of generality (principle of level analysis). The levels up to $n-1$ and those above $n+4$ are not quoted; n is any number.(© 2007, American Scientific Publishers.)

on different levels (levels of reality). According to what principle are these levels of reality arranged? The levels can be arranged vertically, in accordance with the degree of generality where the level with a higher degree of generality is arranged above that with a lower degree of generality (see Fig. 12).

It is obvious that the description of certain facts that are located at a certain level of reality cannot be done by means of another level of reality that is located below this level: a more general structure or theory cannot be deduced on the basis of a structure or theory that is less general (example: quantum theory cannot be deduced from classical mechanics). In other words, the theoretical structures at a certain level cannot be deduced from a lower level.

If we apply these principles on superhuman and human intelligence, we may conclude that a superhuman intelligence cannot be described on the basis of the present human brain structures because the level is by definition below the level of superhuman intelligence. Then, it is obvious that certain experimental brain manipulations can have a result which is in principle not predictable. In other words, such manipulations would be risky and, in particular, irresponsible.

Moreover, there must be no connection between different (brain) levels, because different pictures of reality (world views) can be incommensurable, leading to a completely different assessment of the world outside.

Living systems described mathematically

There is no doubt that the production of superhuman intelligences is dangerous, because such a step can probably not be taken under controlled conditions; principal questions come into play which we have mentioned above. But even when we work only at one level, we get problems: a *complete* description of the present human brain structure is obviously not possible when we apply Gödel's theorem.[11] Let us briefly discuss this fundamental theorem again and let us ask for the consequences regarding the human brain functions and specific gene manipulations.

Within the frame of such a discussion we have always to consider that biological systems (for example a human being) never have the basic, objective reality in front of and around them. It is in any case a species-dependent reality. This point has already been considered in Chap. 1, Sec. 1.4 and Sec. 2.3.2.

Application of Gödel's Theorem

In connection with the controlled manipulation of biological systems, we may state the following: To keep the situation under control, we have to develop theories for the judgment of world views and the description of their changes as a function of specific gene manipulations, i.e. a world view at a higher level. Is that possible at all? Yes, that is in principle possible, but probably not in connection with a biological system that tries to make statements about itself. Here Gödel's theorem comes into play (see also Chap. 1, Sec. 1.4.2).

If we try to formulate such a theory mathematically on the basis of theoretical physics, a living system becomes a mathematical system, which fulfills certain physical principles. To what extent such mathematical systems can be considered adequate remains an open question in the first place.

Then the question "What is the world view of an individual and how is it described?" is equivalent to the question "Can a mathematical system make statements about itself?", and this is because the individual wants to make statements about itself. It is an attempt at self-related analysis.

As mentioned, if an individual wants to make statements with a mathematical system about itself, Gödel's theorem is relevant. Gödel showed[7] that no mathematical system can be complete and consistent simultaneously. A consistent system will always contain statements which one cannot decide whether are true or not. On the other hand, a system cannot be consistent if it contains the complete truth. In mathematics, consistent systems are relevant.

Let us assume that we have a consistent mathematical system (a system is inconsistent if it contains wrong statements besides true ones); so, in accordance with Gödel's theorem, there will be true statements which are not included within the system. The system is missing something, and this is because there are statements which cannot be proven to be true or wrong with system-specific axioms and rules. The reason why the system is incomplete is not that certain information is not yet known, but that completeness is essentially not achievable: we are confronted with the problems which appear if one wants to make statements with a (mathematical) system about the (mathematical) system itself.

In conclusion, the question "Can we develop theories for the judgement of world views and the description of their changes as a function of specific gene manipulations?" must be answered negatively, or at least with care. A biological system that tries to make statements about itself is immediately confronted with Gödel's theorem, provided that this theorem is applicable, i.e. the individual can be described mathematically sufficiently well on the basis of theoretical physics. However, within physics all serious investigations are made in the frame of mathematical statements and, therefore, Gödel's theorem should also be applicable to the theoretical description of physical pictures of the world and their changes due to the manipulation of brain functions.

Summary

Do we really know what a brain is, and how it works? In other words, is a human able to understand his own brain? As we have outlined above, here Gödel's theorem comes into play, and it says that no mathematical system can make complete statements about itself.

When we transfer this theorem to the brain situation, we come to the conclusion that a human can never completely understand his own brain. Such a self-related analysis is obviously not completely possible.

We know a lot about the human brain. The problem is that we do not know what we do *not* know about it. In conclusion, we have to be careful when we "play" with the human brain. Due to Gödel's theorem we cannot completely know what we do when we change the brain functions.

2.3.3 What Kind of Reality Do We Observe?

Gödel's theorem is without any doubt an important principle. But there are further points that we should know when we try to change brain functions. What kind of things are changed when we manipulate brain functions? Here the following question is of particular relevance: Is a human being able to recognize basic, objective reality? What kind of reality is in front of and around him? As stated above, it is not the basic reality, but that what we have in front of and around us is dependent on the biological structure of the observer. This is important to know, and we have studied this point in detail in Ref. 12. Here specific experiments done within behavior research on the basis of animals are of fundamental relevance. Let us summarize the arguments for this realization.

Experiments with animals

One of the natural enemies of the turkey is the weasel — its deadly enemy. When a weasel approaches the nest of a turkey, the later protects its chicks and defends them against the attacker with violent pecks.

Using this system (consisting of a turkey, it chicks, and a weasel), Wolfgang Schleidt performed some basic and highly interesting experiments. These and similar investigations could be as relevant as certain key experiments that have fundamentally changed the physical world view.

Such physical experiments, for example those in elementary particle physics, have to be done on the basis of a certain physical theory; the physical theory is the motivation for the experimental investigation and it dictates the construction of specific experimental devices. On the other hand, the experiments within behavior research (using, for example, turkeys and weasels) do not need a specific physical theory but dictate in a certain sense how reality and space–time are related to each other.[12–15]

As said, Schleidt did his experiments within the usual behavior research, i.e. he did not use expensive and complicated experimental devices such as those that are customary and necessary in for example elementary particle physics, but rather he worked with more or less everyday methods.

It is known that a turkey sitting on its just-hatched chicks attacks everything which approaches its nest. This is of course not true in the case of one of its chicks which has for any reason left the nest. In order to protect the chick, it will steer back the squeaking little bird with calming calls into the nest. All this seems to be nothing more than normal — a matter of course; indeed, the turkey actually shows almost human behavior.

The fact is, however, that the perception apparatus of turkeys must be quite different from that of humans. This can be demonstrated by means of two simple experiments:

(1) Schleidt blocked the ears of the turkey so that it could not hear anything. After a certain period of pacification, one of its chicks approached the nest and a serious disaster happened: without hesitation the turkey strongly pecked at the chick until it was dead.

The turkey saw its chick approaching but did not identify it at all. Everything that is "unknown" and that approaches this bird's nest is attacked. In this specific experiment the chick is no exception and is considered an enemy.

(2) Schleidt implanted into the body of a stuffed weasel a little loud-speaker which emitted the sound of a squeaking little chick. By means of a hidden device he moved the stuffed weasel up to the nest. Also in this case something happened which was quite unexpected: the turkey saw the weasel coming, but did not identify it; after some hesitation, it even allowed the weasel into the nest and gave it protection.

These dramatic and unexpected results lead to the conclusion that the turkey optically experiences the world quite differently from the way human beings do, even though its eyes are quite similar to those of man. There is obviously no similarity between what the turkey experiences and what a human being sees in the same situation.

All turkeys experience the world in this way. In other words, the experiments are reproducible and demonstrate a general scientific fact, i.e. it is not only an individual instance.

Respectively correct, but not comparable with each other

How can we integrate these facts from behavior research, which no one would have even approximately expected in this form, into an ordered whole? If we presuppose that human beings are further developed in evolution than turkeys, then it seems at first sight to be reasonable to assume that the turkey perceives reality in a more rough-and-ready, less sophisticated way than man does. But this does not account for the drastic changes in the turkey's general perception which are brought about by blocking its ears. For this would not be expected to influence the visual signals it receives and processes. This, however, is precisely the case. Turkeys appear not merely to have a rougher visual perception of reality than humans, but obviously to have a qualitatively different one, which is influenced in particular by acoustic signals. That is to say, turkeys experience the world differently from the way man does.

On the other hand, both systems — man and turkey — react correctly in the normal case because both species are able to exist in the world, which can only be possible from the viewpoint of the modern principles of evolution if their particular views of the world are correct.

Therefore, although the conceptions of the world of man and turkey are on the one hand different from each other, they are on the other hand correct in each case. This means that neither of these two conceptions of the world can be true in the sense that they are a faithful reproduction of nature: Basic, objective reality must be different from the images which biological systems construct of it. This is an important statement.

The assumption that only man is able to recognize true, absolute reality would be arrogant and not be justified, especially as man is subject to the same principles of evolution as other biological systems. We must consider all biological systems as equivalent in principle, even when a human being has reached a relatively advanced stage of development.

There are no scientific reasons for the assumption that man is separate from the whole scenario and therefore occupies a special position among the biological systems. As said, there are no scientific reasons for that.

Thus, if all biological systems are equivalent and, on the other hand, if they have versions of reality in front of them which are in general different from each other, as in the case of man and turkey, neither of the versions can exist in reality outside. If any representation would nevertheless reflect basic, objective reality, this biological system would have an advantage over the others. That to say, not all biological species could be equivalent. But as mentioned, there are no scientific reasons for such an assumption.

Nanotechnological changes of the brain functions

From all these statements in connection with behavior research, we may conclude that the world view of human beings — the perception of the world in front of and around us — is dependent on the biological structure of man, i.e. on the brain functions. However, this world view is essential for individuals for survival, at least in the early phase of evolution, and each phase later is based on the biological structure developed in the earlier phase of evolution.

If we change the brain functions nanotechnologically, we change the world view and possibly disturb the relationship between the human being and his environment, which can lead to a disaster, just as we have learned from the chick experiment within the frame of behavior research.

Even when the new world view — produced by nanotechnological changes of the brain functions — is more powerful and efficient than the old one, the human being is in general no longer adapted to the environment and is also no longer in accord with all the other human beings. To conclude, such a situation can have the effect that such a human being is not able to survive; we cannot explicitly exclude such a terrible scenario.

In conclusion, a human being is obviously not able to recognize what we have called "basic reality"; exactly the same should be true of other species, for example the turkey. Man uses certain information from basic reality for the construction of a species-dependent reality, which is tailor-made for him, and he recognizes this human-dependent reality on various levels — levels of reality. This is just the conception we used in Sec. 2.3.2 and is in particular reflected in Fig. 12.

Question: What is the reason for that? Why do man and other species not observe basic reality but only a certain kind of "second-class reality"? The answer is given by the principles of biological evolution.

As is well known, evolution changed and still changes biological systems continuously in the course of time from one state to another, where the time intervals from one step to another are in general very large. Nature is obviously careful! The adaptation to the environment of the biological system must succeed under all circumstances. Otherwise the species under investigation has no chance to survive and will be rooted out.

In principle, we try to make such changes to the biological structure also in a nanotechnological way, just in analogy with evolution. However, the envisaged nanotechnological manipulations are planned to be more drastic, i.e. the step from one biological state to another will be relatively small in nanotechnology in comparison with natural evolution, and this can be dangerous because the adaptation to the environment can fail. We have to be careful when we try to manipulate members of a species.

Again, why do we "only" observe a species-dependent (second-class) reality, which is by definition species-independent, and not the true, basic reality? This question is equivalent to "What are the principles of evolution?" Because this is an important point also in connection with nanotechnological manipulations, let us briefly discuss in the next section the main peculiarity of evolutionary processes, in particular what we can call the "strategy of nature."

2.3.4 The Strategy of Nature

It is conceivable that conceptions of the world (world views) that are different from each other can be equivalent concerning their information and efficiency. This is not a contradiction because world views of different biological systems, which cannot be compared with each other, could be equivalent.

In accordance with the discussion above, more highly developed organisms, for example an organism which could be descended from the turkey, could have developed epistemologically equivalent conceptions (world views) in comparison with man, although the conception of one system is different from that of the other. In particular, both biological structures could have developed a reality in front of and around them which are different from each other.

These world views, developed by different biological structures, are not *wrong*, but they are also not *true* in the sense of a precise reproduction of the true, basic reality in the way that a photograph represents a one-to-one copy of the world. That the features in the worlds in front of the various species cannot be wrong, or true in the sense of a precise reproduction of basic reality, can already be extracted from the strategy on which nature is based, which is reflected in the principles of evolution.

An important principle: "as little outside world as possible"

The perception of true reality in the sense of a precise reproduction implies that with growing fine structure in the picture, increasing information of actual reality outside is needed. Then, evolution would have developed sense organs with the property of transmiting as much information from reality outside as possible. But the opposite is the case: the strategy of nature is to take up as little information from the outside world as possible. Reality outside is not assessed by "true" and "untrue" but by "favorable to life" and "hostile to life."

This is probably the most important idea with respect to the phenomenon of evolution. Concerning this remarkable statement, Hoimar von Ditfurth formulated this important fact by the following statement[16]: "No doubt, the rule 'As little outside world as possible, only as much as is absolutely necessary' is apparent in evolution. It is valid for all descendants of the primeval cell and therefore also for ourselves. Without doubt, the horizon of the properties of the tangible environment has been extended more and more in the course of time. But in principle only those qualities of the outside world are accessible to our perception apparatus which, in the meantime, we need as living organism in our stage of development. Also our brain evolved not as an organ to understand the world but an organ to survive."

As said, the principle "as little outside world as possible" can be understood by means of the idea of evolution, which is briefly discussed in the next section.

Some principle remarks on biological evolution

The principles of evolution, i.e. the phylogenetic development from simple, primeval forms to highly developed organisms, can be considered as the key to the "perception of reality" of biological systems, such as man and the turkey. It is the theory of evolution by natural selection which is generally accepted in the meantime; its foundations have been created by Charles Darwin more than 100 years ago. Since then it has been modified and developed further by geneticists.

Evolution by natural selection is a two-step process. *Step 1*: By recombination, mutation, etc., genetic variants are produced at random. Populations with thousands or millions of independent individuals arise. *Step 2*: Some of these independent individuals will have genes which enable them to manage the predominating situation due to the environment (climate, competition, enemies), better than others. They thus have a larger chance of survival than others; they will have, on the statistical average, more descendants than other members of the population.

Natural selection takes place in favor of those organisms whose genes have adapted to better cope with the environment. The number of examples that biological structures have developed in accordance with these two criteria is overwhelming. In connection with our discussion, it is important to mention that the species-dependent world views formed by the different biological structures (such as man and the turkey) must also be characterized by this species-preserving appropriateness.

Not cognition but the differentiation between "favorable to survival" and "hostile to survival"

To conclude, in nature it is not cognition that plays the essential role but the differentiation between "favorable to survival" and "hostile to survival," at least in the early phases of evolution. With respect to this fundamental point, it is important that the members of a species are able to recognize and to assess the earliest possible changes in the environment within the framework of a consistent world view, in particular within the reality in front of and around a member.

The possibility of a fallacy has to be excluded as early as possible. There may be no doubts concerning the particular status of the environment. For this purpose, the reality in front of a biological system, which has been designed through evolutionary processes, has to be correct but it may contain, for economic reasons, only those things which are absolutely necessary for survival. Everything else is unnecessary; it encumbers and is therefore hostile to life.

The species-dependent world view does not to have to be complete and true (in the sense of a precise reproduction with respect to basic, species-independent reality), but it must be reliable and restricted, at least during the early phylogenetical phase. These are the criteria that guarantee optimal chances of survival.

Nanotechnological manipulations

All that has to be kept in mind when we plan nanotechnological changes within the frame of biological systems, in particular when we try to manipulate specific brain functions. Since the world views are species-dependent and extremely relevant to survival, we change in general also the world view when we manipulate brain functions nanotechnologically. In other words, in such cases we change everything that has been developed by evolution during the entire past, and not only what has been developed in the last phases. That is to say, we have to be careful because we deal here with a complex situation.

Summary

From the point of view of evolution we may state that the impressions in front us do not reflect basic, objective reality, which exists independently of the observer. But what we see in front of and around us reflects "merely" an appropriate species-dependent reality that is formed by the individuals from certain information from the outside world.

These species-dependent realities are incomplete; they are correct but not necessarily true. The last point follows directly from the fact that different organisms, such as man and the turkey, form world views which are distinct from each other. None of them can be wrong, because the concerned species would inevitably die out in the case of a wrong world view.

All these facts can be extracted from the strategy of nature which is summarized in the principles of evolution. After that, it is of primary importance that we can reliably distinguish between "favorable to survival" and "hostile to survival," and not to form a true world view. An individual registers situations in the environment by certain patterns, which are tailor-made for the particular needs of the species and relieved of the condition to be a precise reproduction of the true, basic reality.

So, for example, in order to find a certain place in a cinema, it is not necessary that a visitor gets at the pay desk a small but true model of the cinema, i.e. a precise reproduction of the cinema, which is reduced in size; a simple cinema ticket with the essential information, where to go, is more appropriate. In this respect, the cinema ticket is what we have called above species-dependent reality; the cinema itself or the true, small model of it reflects basic, objective reality. An individual needs only such information of basic reality that is necessary for certain purposes.

Thus, a human being does not have to know a precise reproduction of basic reality. There must be no one-to-one correspondence between what an observer actually sees — which is in front of him — and what exists independently of him, i.e. basic reality. There is no one-to-one correspondence between the cinema ticket and the real cinema. The species-dependent reality (here the cinema ticket) has to be correct but it may only contain, for economic reasons, information which is absolutely necessary for finding the place in the cinema. Everything else in unnecessary.

All that means the following: we do not know how reality is actually constructed, i.e. how it is constructed independently of the observer's perception apparatus. It is probable that we will never know that, because we are caught by our own cognition system and a species-independent point of view is not conceivable. Therefore, basic, objective reality remains essentially hidden.

Lashley and Pribham

In fact, specific experiments with animals show convincingly that there should be no one-to-one correspondence between the image in the brain and the real structure in the outside world; it is obviously not a photograph-like image. This is without doubt surprising and is against the standard understanding in physics, etc. In Ref. 18 we find the following important comment:

"Another of Lashley's discoveries was that the visual centres of the brain were also surprisingly resistant to surgical excision. Even after removing as much of 90 per cent of a rat's visual cortex (the part of the brain that receives and interprets what the eye sees), he found it could still perform tasks requiring complex visual skills. Similarly, research conducted by Pribham revealed that as much as 98 per cent of a cat's optic nerves can be severed without seriously impairing its ability to perform complex visual tasks."

"Such a situation was tantamount to believing that a movie audience could still enjoy a motion picture even after 90 per cent of the movie screen was missing, and his experiments presented once again a serious challenge to the standard understanding of how vision works. According to the leading theory of the day, there was a one-to-one correspondence between the image the eye sees and the way that image is presented in the brain. In other words, we look at a square, it was believed the electrical activity in our visual cortex also possesses the form of a square..."

"Although findings such as Lashley's seemed to deal a death blow to this idea, Pribham was not satisfied. While he was at Yale he devised a series of experiments to resolve the matter and spent seven years carefully measuring the electrical activity in the brain of monkeys while they performed various visual tasks. He discovered that not only there's no such one-to-one correspondence but there wasn't even a discernible pattern to the sequence in which electrodes fired. He wrote of his findings: These experimental results are incompatible with a view that a photograph-like image becomes projected onto the cortex surface."

We may state the following: The chick experiment revealed quite clearly that the visual impressions in front of and around us are species-dependent. Lashley's and Pribham's discoveries underline this statement distinctly. Animals could still perform complex tasks without having the entire picture of outside reality in front of them, and it is not a tiny fraction of the information in this picture that is absent but 90%. It is obvious that in such cases the brain functions are particularly active; the brain functions are always active, but in such cases exceptionally so. Science knows that within all visual perceptions in front of and around us, the eyes, the optic nerves, and the brain work together.

Von Foerster

That the outside world (basic, objective reality) cannot be of the composition that we feel it has, becomes particularly apparent when we realize that there is, for example, no light and no color in the outside world. Since the world before us (before our eyes) is essentially formed by light and colors, this world cannot be the outside world; what we call the outside world exists only in the head of the observer. In this connection, we find in Ref. 18 the following relevant comment:

"The eminent cyberneticist Heinz Von Foerster points out that the human mind does not perceive what is 'there', but what he believes should be there. We are able to see because our retinas absorb light from the outside world and convey the signals to the brain. The same is true of all our sensory receptors. However, our retinas don't see color. They are 'blind', as Von Foerster puts in, to the quality of their stimulation and are responsive only to their quantity. He states, 'This should not come as a surprise, for indeed out there there is no light and color, there are only electromagnetic waves; "out there" there is no sound and no music, there are only periodic variations of air pressure; "out there" there is no heat and no cold, there are only moving molecules with more or less mean kinetic energy, and so on. Finally, for sure, "out there" there is no pain. Since the physical nature of the stimulus — its quality — is not encoded into nervous activity, the fundamental question arises as to how does our brain conjure up the tremendous variety of colors as we experience it at any moment while awake, and sometimes in dreams while asleep.'"

This statement by Von Foerster should be as basic as those formulated by Lashley and Pribham. The standard understanding that the world in our brains is a one-to-one representation of what is positioned in basic, objective reality is obviously a fallacy and is no longer tenable. We never perceive basic reality but it is always a species-dependent world view (point of view), which particularly is dependent on the brain functions of the observer.

This is of course extremely important for nanotechnological manipulations with the brain, because we change in general the world view of a species and, therefore, we could change the imagination of what the biological system needs for survival. Here we have to be careful, be very careful.

Even when we invent nanotechnological methods to enhance the intelligence of a human being, these changes could in principle modify simultaneously the world view of the observer that he needs for survival. In Sec. 2.3.1 we have discussed a specific nanotechnological experiment for increasing mousy intelligence, and similar experiments are explicitly planned for man. This is interesting, but what do we change simultaneously? Such and similar questions have to be investigated with the help of reliable theoretical models before we start such experiments. Are the present theoretical tools really sufficient for an adequate description?

The world we experience is an invention of the brain

We have learned from the statement by Von Foerster quoted above that light, as well as other things, does not exist in reality outside (basic reality). Yet, this is exactly what people need most. The "world" which our brain makes of reality (as we have discussed above, it is the species-dependent world) is mostly an invention of the brain; we have seen within the frame of the chick experiment that this invention obviously depends strongly on the biological species, since the turkey obviously had a quite different visual perception of a chick than a human being.

2.3.5 Scientific Realism

The naive point of view

As is well known, the common or naïve point of view assumes that the inner picture, which directly appears in front us, is identical with the events outside us. That is to say, it is held that natural phenomena exist doubly, so to speak. For example, the famous psychologist C. G. Jung formulated this fact as follows[19]: "When one reflects upon what consciousness really is, one is profoundly impressed by the extreme wonder of the fact that an event taking place outside in the cosmos simultaneously produces an internal image, that it takes place, so to speak, inside as well, which is to say: becomes conscious."

According to Jung, we therefore have a reality and a picture of it; an event outside is simultaneously an inner event. This is of course not new to us, but Jung assumed that the two events are identical, that the structure in the material world outside is identical with that in the picture. However, the chick experiment and in particular the investigations by Lashley, Pribham, and Von Foerster teach us that the structures outside should not be identical with the structure in the picture.

When we try to manipulate nanotechnologically the brain of a biological system (man, turkey, etc.), the world view of the corresponding species must be preserved, and it is extremely important to know that the information in the world outside is actually not transferred as one-to-one information into the innerpart of biological systems.

From the debate in the last section, which is essentially based on the investigations by Lashley, Pribham, and Von Foerster, it follows that the structures and characteristics in the world in front of us (it is a picture) cannot be identical with those in basic reality. However, not only Jung but most people and also scientists believe the opposite: The structures in the directly perceived image are identical with those in the world outside, although we have no access to basic, species-independent reality because we are caught by our own cognition system and a species-independent point of view is not conceivable; we have discussed this point already. Nevertheless, it is almost always assumed that the two types of reality are identical.

As said, most people assume automatically that the things in front of them are the material world itself and, as remarked, those humans who are conscious of the fact that it is only a picture normally assume as a matter of course that the structures and the other characteristics (for example the color) in the picture are identical with those in the objective material reality.

It is evident that there are no material objects in the picture but only geometrical structures, and that what we see in front of us is a "picture of reality" and not the material reality itself; we have designated above the pictures as "species-dependent reality." However, as has been mentioned several times, most people assume that the structures in the picture are identical with those in the actual reality outside. Even scientists believe that and simply ignore the basic investigations by Lashley, Pribham, and Von Foerster.

Let us consider this point in more detail. Two objects (for example two planets) appear in the picture in front of us as two geometrical points having the position 1 and 2, and these geometrical positions are symbolized by crosses in Fig. 13; r_{12} is the distance between 1 and 2. We assume in the most natural way that the composition in the picture (Fig. 13) is identical with the situation in reality, i.e. we assume that in reality the geometrical positions in the picture are replaced by material objects (Fig. 14), where the full points A and B represent the material objects in the actual reality outside; the distance between A and B is given by $r_{AB} = r_{12}$, because the geometrical positions (crosses) are merely replaced by the material objects.

But is this realistic? It is definitely not, as we have recognized above with a large number of indications. How is the world structured in reality outside? This question is answered in the sections below: we cannot say anything about the structures outside because

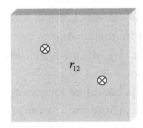

Fig. 13 Two material objects for example two plants, appear in front of us as geometrical positions; r_{12} is the distance between these two positions.

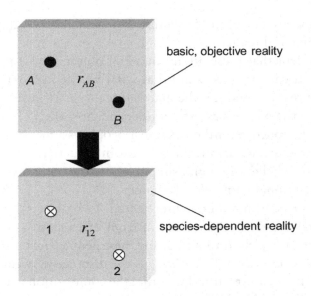

Fig. 14   The geometrical positions in the species-dependent reality (crosses) are replaced by material objects in the basic, objective reality, where the full points A and B represent the material objects in the actual reality outside; the distance between A and B is given by $r_{AB} = r_{12}$, because the geometrical positions are merely replaced by the material objects. However, this is a fallacy, because the species-dependent reality is not identical with the actual reality outside.

a species-independent point of view does not exist and is not conceivable, and we come to Fig. 15.

The role of the equations of motion

What does the realism mean that is reflected by Fig. 14? Let us briefly discuss this point by classical mechanics. Classical mechanics claims to describe the motion of material objects as it really takes place in the actual reality outside (Fig. 14, upper part). Here Newton's equations of motion are responsible.

Newton's mechanics is based on observations which the observer makes in everyday life, i.e. these are observations on the basis of pictures that are directly in front of the observer (Fig. 13 and lower part of Fig. 14). Newton's equations of motion describe material objects on their way through space and time. However, in the picture, it is not

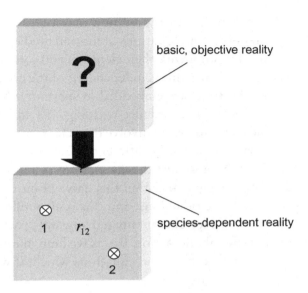

basic, objective reality

species-dependent reality

1   r_{12}

2

Fig. 15   The geometrical positions in the species-dependent reality (crosses), which just represent the world before and around us. How basic, objective reality is formed cannot be recognized by a human observer, since he cannot take a species-independent point of view. This is symbolized by a question mark.

that the material objects are moving relative to each other, but that the geometrical positions are moving. This fact is then unimportant for the equations of motion when we assume that the composition of the actual reality outside is identical with the structure in the picture, i.e. when the scheme in Fig. 14 is valid.

Newton himself certainly assumed that we are "embedded in space," whereas the scheme in Fig. 14 indicates that it should be a projection onto space, i.e. reality is projected onto space. That is a difference in principle — which is, however, not relevant if the structures in the picture are identical with those in reality outside.

However, Fig. 14 reflects a scientific realism which, in our opinion, is questionable; such a projection would be, for the following reason, too naïve: within the framework of Newton's theory the masses and the interactions between them are considered to be objectively real. Also, the solutions to Newton's equations of motion, i.e. the trajectories of the masses, are considered to be objectively real because we observe with high precision the motion of a celestial body (for example the moon), just as the equations of motion predict. In conclusion, the equations of motion with their elements and solutions reflect how reality is composed, and this is independent of the conception of whether the material objects are embedded in space or whether they are projected onto space (in accordance with Figs. 13 and 14).

Such a scientific scenario would involve the masses solving, in their motion through space, incessantly differential equations (Newton's equations of motion), because just these equations of motion have to be considered as objectively real within this naïve point of view. But such a realism seems to be ridiculous, and this is also reflected in the following remark[20]: "As Herschel ruminated long ago, particles moving in mutual gravitational interaction are, as we human investigators see them, for ever solving differential equations which, if written out in full, might circle the earth."

If we use Figs. 13 and 14 as a basis, the equations of motion refer primarily to the elements in the picture (as has been discussed above, we have in any case only pictures in front of us and not reality itself). But this does not mean that nature (here the masses) solves incessantly differential equations, because the picture is a construction of the biological cognition apparatus, just as the equations of motion are the result of a cognition process.

If we use the equations of motion only for the description of the elements in the picture, the conception that the masses solve, in their motion through space, incessantly differential equations is avoided. Then, we have to conclude that Fig. 14 cannot be correct. The reason is that we have to also assume here that there is in general no similarity between the structures and characteristics in the picture (Fig. 13 and the lower part of Fig. 14) and those in the actual reality outside.

Here also, how the actual reality is constructed cannot in principle be said (Fig. 15 is valid), because there is not a picture-independent point of view for the observer. We have already stated above that the reality in front of biological systems is species-dependent.

In other words, in connection with the equations of motion we come to the exact same conclusion that we have drawn above from the principles of evolution, which are in particular supported by the specific and realistic investigations of Lashley, Pribham, and Von Foerster.

In conclusion, the world is actually not as science takes it to be, and the furnishings are not as present science envisages them to be. Rescher studied the structures of scientific laws carefully, and he remarked[20]:

"Scientific realism is the doctrine that science describes the real world that the world actually is as science takes it to be, and that its furnishings are as science envisages them to be. If we want to know about the existence and the nature of heavy water or quarks, of man-eating molluscs or a luminiferous aether, we are referred to the natural sciences for the answer. On this realistic construction of scientific theorizing, the theoretical terms of natural science refer to real physical entities and describe their attributes and components. For example, the 'electron spin' of atomic physics refers to a behavioural characteristic of a real, albeit unobservable, object — an electron. According to this currently fashionable theory, the declarations of science are — or will eventually become — factually true generalizations about the actual behaviour of objects that exist in the world. Is this 'convergent realism' a tenable position?"

"It is clear that it is not. There is clearly unsufficient warrant for and little plausibility to the claim that the world is indeed as our science claims it to be — that we've got matters altogether right, so that our science is correct science and offers the definite 'last word' on the issues. We really cannot reasonably suppose that science as it now stands affords the real truth as regards its creations of theory."

This comment by Rescher shows, together with the above discussion in connection with the role of the equations of motion, that we have to firmly assume that the scientific pictures have only limited similarity to what actually takes place in reality outside.

Since, on the other hand, the scientific pictures are essentially based on our observations in everyday life (in particular, they have to be compatible with them), we may also conclude that those things and processes which we have in front of us in everyday life can also have only limited similarity to what actually takes place in reality outside. Von Foerster's statement in Sec. 2.3.4 supports this point of view.

Conclusion

In the development of biological structures and their changes, nano-technology and nanoscience will belong to the most relevant future disciplines. Here, not only the *world view* of an individual is relevant — in particular the question whether or not this world view is species-dependent — but also the question of the underlying scientific scenario (order) behind all these things. Is it, for example, the scientific realism discussed in this section, where the particles incessantly solve the equations of motion, etc., or is it the conception of species-dependent realities? We came to the conclusion that a species-dependent conception is more realistic because it correlates a relatively large set of experimental observations, for example the chick experiment.

In the case of the scientific realism outlined in this section (the world outside is actually composed as science takes it to be, i.e. science describes the real world outside), the world views for all biological systems (man, turkey, etc.) would be the same. Without doubt, such a naïve conception can lead to problems when we try to manipulate nanotechnologically the brain functions and/or other biological mechanisms. However, our present standard understanding, on which we base nanotechnology, is just given by this naïve conception. Without doubt, this can lead to wrong assessments.

Consequences for physics

We develop our new inventions in nanoscience and nanotechnology on the basis of the principles of physics. Do we have we to modify these principles in the future?

Physics is thought to be able to make the most exact statements about the composition of reality, as it is recognized as the basic science. But if that which we perceive does not exist at all in reality outside, then we must ask ourselves of what physics can make statements about at all. This is an important point when we manipulate the brain functions of biological systems within nanotechnology.

Must we take the sun, moon, and stars as metaphysical objects if they do not exist in the forms we have in front of us, which are, in other words, directly perceived by us? This question not only arises in the case of the sun, moon, and stars, but concerns everything which we mean by so-called "hard" objects, also regarding man himself: the visual image which man has of himself does not exist in that form in the actual reality outside. The form in which we perceive man and all the other things has no material counterpart.

This conclusion is actually more extraordinary than the results revealed by the chick experiment. The sun, moon, and stars and all the material objects should nevertheless not be considered as metaphysical elements, because they appear in the picture (the species-dependent reality) as transformed objects. How all these things appear in absolute reality (reality outside) cannot be assessed, because the human observer is essentially not able to take a picture-independent point of view and we come to Fig. 15; evolution developed human beings and other biological systems on the basis of other criteria.

2.3.6 Kant's Philosophy

The philosopher Immanuel Kant investigated the relation between true and perceived reality. His ideas are close to what we have developed above (Secs. 2.3.1–2.3.5). However, there are nevertheless big differences concerning statements about reality outside. Let us briefly discuss Kant's ideas.

Kant argued that we cannot make statements about the true reality outside. According to him, all things we observe are located within space–time and these elements, space and time, are located inside the observer. In his opinion, a human observer can say nothing about the structure of the outside world. In particular, in his opinion a human observer is *not* able to give answers to the following questions: Is there a one-to-one correspondence between the structures outside and those in the brain of the observer? Does the information in the pictures located in the brain reflect the complete information about reality outside? This is because Kant knew nothing about the principles of evolution, but this is an essential point in answering such questions.

According to Kant, space and time are not empirical concepts, which are determined by abstraction from experience. Experiences become possible at all only through the concepts of space and time. According to Kant, space and time are not objects, but have to be considered as preconditions for the possibility of all experiences. Although in his opinion space and time are not empirical concepts, they nevertheless have empirical reality. This is because all things which we observe are located in space and time. The structures of space and time are therefore reflected in the empirical objects. Kant denied the existence of a space and a time independent of brain functions (observations in everyday life and thinking).

According to Kant, space and time are located inside the observer. Whether space and time are also elements of the actual (fundamental) reality outside remains essentially an open question within his point of view.

Kant's perspective is without any doubt important, not only in connection with philosophical questions. But what are the consequences for physics? If we take Kant's view seriously, then the physical laws, for example Newton's gravitational law, are merely pictures in the head of the observer and there is essentially no way to express it for the reality outside; nothing can be said about the processes in the outside world, and this is consistent with Fig. 15.

This is in contrast to "projection theory" (outlined in Refs. 13–15), where we can construct fictitious realities. But also within projection theory space and time are auxiliary elements for the description of the world outside.

If the gravitational law (and all the other physical laws) is merely a picture in the head, we get a problem, because there can be no gravitational forces in the head of the observer. In conclusion, Kant's thoughts can lead to considerable problems when we apply them to physics. Barrow remarked[21]: "We can see that Kant's perspective is worrying for the scientific view of the world." However, Kant's perspective has not been taken so seriously in physics and there are other positions. The situation was well analyzed by Barrow: "There are two poles about the relationship between true reality and perceived reality. At one extreme, we find 'realists', who regard the filtering of information about the world by mental categories to be a harmless complication that has no significant effect upon the character of the true reality 'out there'. Even when it makes a big difference, we can often understand enough about the cognitive processes involved to recognize when they are being biased, and make some appropriate correction. At the other extreme, we find 'anti-realists', who would deny us any knowledge of that elusive true reality at all. In between these two extremes, you will find a spectrum of compromise positions extensive enough to fill any philosopher's library: each apportions a different weight to the distortion of true reality by our senses."[21]

In other words, there are no criteria for deciding about the true nature of absolute reality; the realists cannot disprove the antirealists and vice versa. The realists more or less assume that there is a one-to-one correspondence between true reality and perceived reality (picture); the antirealists maintain that we can say nothing about true reality.

However, when we consider the basic facts of biological evolution, both viewpoints seem to be unrealistic. Evolution teaches us that for humans and animals it is not cognition that plays the important role in nature but the differentiation between "favorable to survival" and "hostile to survival," at least in the early phase of evolution (Secs. 2.3.3 and 2.3.4). Each picture of reality (perceived reality) is tailor-made for these characteristics. Since the conditions for survival are different for different biological systems, the perceived realities are also different for different biological systems. Wolfgang Schleidt's experiments (Sec. 2.3.3) with a turkey showed that very impressively.

The picture of reality designed by the individual (unconsciously) has to be correct but it may only contain, for economic reasons, information which is absolutely necessary for survival; everything else is unnecessary.

The picture of reality does not have to be complete and true (in the sense of a precise reproduction) but, rather, restricted and reliable. Furthermore, we learned from Schleidt's experiments (Sec. 2.3.3) that the conception of the world of man and that of the world of the turkey are on the one hand different from each other, and on the other hand correct in each case. In other words, neither of these two conceptions of the world can be true in the sense that they are faithful reproductions of nature.

This means that there can be no one-to-one correspondence between the structures in the picture and those in true reality. In other words, Fig. 14 cannot reflect the true situation. Objective reality must be different from the images which biological systems construct from it, but we cannot know the structure of true, basic reality (Fig. 15); we cannot even estimate it.

The statement that there can be no one-to-one correspondence is, on the one hand, against the position of the realists and, on the other hand, simultaneously against the position of antirealists because it is a statement about true reality. Here also, the position that is expressed by Fig. 15 is more realistic.

2.3.7 Experiments with Distorting Glasses

The entire conception regarding basic, objective reality and species-dependent realities is supported by certain experiments: the so-called experiments with distorting glasses.

If one uses glasses which strongly distort the picture in space, the following takes place: after a certain time the observer sees everything in the normal order, i.e. the spatial situation is the same as before without distorted glasses, and space is again right-angled (Euclidean). In other words, the distorting glasses are effectively ignored after a certain time by the observer's cognition apparatus.

In conclusion, distorting glasses transform the properties of space (its metric), which satisfy the axioms of Euclidean geometry, into a space with non-Euclidean geometry. Without changing the physical conditions, we observe after a certain time the following strange effect: the cognition apparatus of the observer obviously transforms space with non-Euclidean geometry into the usual space, which satisfies the axioms of Euclidean geometry. In other words, the cognition apparatus of the observer is able to influence the space and the picture, respectively.

Again, after a certain time the non-Euclidean metric transforms back and we get again the picture with Euclidean geometry and with straight rays of light, which we observe normally in experiments. The brain simply ignores the distorting glasses. It is probably better to say that the brain goes back to Euclidean processing.

The real physical process outside remains unchanged; only the kind of perception has been changed. We have two geometries for one and the same process outside. The brain processes temporarily on the basis of non-Euclidean geometry.

Without doubt, the experiment with distorting glasses is basic with respect to the nature of space and the role of the observer. In particular, it demonstrates convincingly that the world in front of us is actually inside the head.

What does that mean for nanotechnology? We could change the brain functions permanently in such a way that the geometry becomes non-Euclidean. In this case the brain will not transform the metric back to a space with Euclidean geometry. This is an interesting point, and we come to the findings of D'Arcy Wentworth Thompson.

2.3.8 D'Arcy Wentworth Thompson

The biologist D'Arcy Wentworth Thompson (1860–1948) investigated changes in biological organisms by means of geometrical considerations. He made important discoveries and was able to describe systematic changes in biological organisms that occur in the course of evolutionary developments, by means of changes in geometry (its metric). His pioneering discovery was that the changes during these evolutionary developments can be described by a change in geometry (the metric), in most cases from Euclidean to non-Euclidean geometry. In this way the pictures of one living organism could be transformed into others. Thompson's famous book *On Growth and*

Form, published in 1917, has had a considerable echo in the scientific community.

Thompson's kind of analysis (see also Ref. 22) can be applied to many organisms and many biological details concerning bones, etc. In particular, he threw light on the role of geometry in connection with physically real processes. The discovery by Thompson is therefore of particular relevance to what we have analyzed in Ref. 13: the nature of space and time. Here we will make only some general remarks on his discovery, and we will apply that to fish.

Fish

Although our cognition apparatus works on the basis of Euclidean geometry, we have recognized that there are alternatives in the form of curved, non-Euclidean spaces. Any pattern drawn on a blank piece of paper can be connected with a curved or Euclidean coordinate system. It is a matter of convention.

Figures 16 and 17 show fishes (*Diodon*, *Orthagoriscus*, *Scarus sp.*, and *Pomacantus*) in space, i.e. organisms as we experience them in everyday life, and these experiences occur in a Euclidean space.

Systematic changes from *Diodon* to *Orthagoriscus* and from *Scarus sp.* to *Pomacantus* occur in the course of evolutionary processes. Thompson explained the evolutionary changes geometrically, i.e. he changed the geometry (metric) in order to go from *Diodon* to *Orthagoriscus* (Fig. 18) and from *Scarus sp.* to *Pomacantus* (Fig. 19). In other words, he applied to *Diodon* and *Scarus sp.* non-Euclidean geometry and got *Orthagoriscus* and *Pomacantus*.

Application to nanoscience

In principle, we were able to change nanotechnologically the geometry (metric) within the brain, produced by the brain functions, in such a way that we obtained, for example, the fish *Orthagoriscus* within non-Euclidean geometry (Fig. 18), i.e. we obtained, instead of *Diodon* within Euclidean geometry (produced without nanotechnologal impact), *Orthagoriscus*.

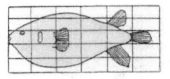

Diodon

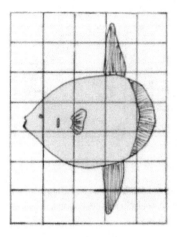

Orthagoriscus

Fig. 16 Two fishes (*Diodon* and *Orthagoriscus*) as we experience them within the observations in everyday life (within Euclidean geometry). *Diodon* gradually changes within an evolutionary process, resulting in *Orthagoriscus*. Both fishes really exist in nature.

In other words, in reality outside is the fish *Diodon* but the image of it in front of us (within species-dependent reality) is the fish *Orthagoriscus*, and this is entirely due to the nanotechnological change of the brain functions. Since *Diodon* and *Orthagoriscus* are different from each other, they are underlying different living conditions (at least in principle). If the observer knows nothing about the change (transformation) within his brain, he treats *Diodon* as *Orthagoriscus*. This particularly means that the treatment based on the picture is no longer compatible with what is in the actual reality outside. This can lead to problems.

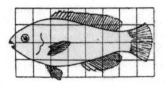

Scarus sp

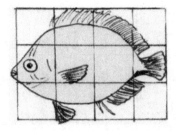

Pomacanthus

Fig. 17   Two fishes (*Scarus sp.* and *Pomacantus*) as we experience them within the observations in everyday life (within Euclidean geometry). *Scarus sp.* gradually changes within an evolutionary process, resulting in *Pomacantus.* Both fishes really exist in nature.

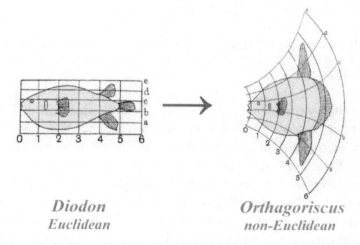

Diodon
Euclidean

Orthagoriscus
non-Euclidean

Fig. 18   Thompson transition: *Diodon* ⟶ *Orthagoriscus.* Transition from Euclidean to non-Euclidean geometry. (Reprinted with permission from Ref. 22; © 1992, Dover, New York.)

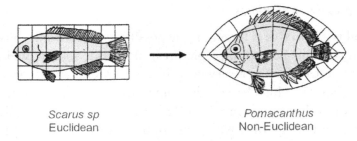

Scarus sp
Euclidean

Pomacanthus
Non-Euclidean

Fig. 19   Thompson transition: *Scarus sp.* ⟶ *Pomacantus*. Transition from Euclidean to non-Euclidean geometry. (Reprinted with permission from Ref. 22; © 1992, Dover, New York.)

Clearly, exactly the same is the case of *Scarus sp.* and *Pomacantus* (Fig. 19). Here the observer treats *Scarus sp.* as *Pomacantus*. Also in this case, compatibility is no longer a given and is violated.

We can generalize that, and we have to assume that, any change in the brain functions can lead to world views that are no longer compatible with what is really outside. The reason is obvious, because changes in the brain functions transform the image in front of us (the species-dependent reality) but not the actual reality outside. Normally, we assume that when the image in front of us is changed, reality itself is changed. We have to careful — probably very careful.

2.3.9 Compatibility: Some Principal Remarks

The solution of specific problems

The pictures that appear in species-dependent reality must be compatible with what is outside. This is an important point. Let us therefore make some principal remarks on it.

Man and other biological systems have developed so that they are able to solve specific problems consistently and conclusively. Such a principle is necessary for survival. This is valid for the conscious cognition apparatus as well as for the unconscious one, and of course for the anatomical setup.

Species-preserving principles

The ability to survive is based on the possibility of forming conscious and unconscious perceptions (pictures) of outside reality, which are tailor-made to species-preserving principles. Any picture consists of certain elements which have to be compatible with each other.

On the other hand, the elements in pictures of one species are in general not compatible with those of other species. Each biological system forms a separate system. In other words, the picture elements of different species are in general not interchangeable. If one tries it nevertheless, difficulties can appear, as in the case of the chick experiment by Schleidt (Sec. 2.3.3); he tried to project the picture elements, which are characteristic of human behavior, onto the perception apparatus of the turkey, and the result was a disaster.

Exactly the same is the case of *Diodon* and *Orthagoriscus* (Fig. 18), as well as *Scarus sp.* and *Pomacantus* (Fig. 19). Here the observer treats *Diodon* as *Orthagoriscus* and, on the other hand, *Scarus sp.* as *Pomacantus*. Also in these cases, compatibility is in general no longer a given and is possibly violated. This particularly means that the treatment based on the picture is no longer compatible with what actual happens in reality outside.

Petrol in an electric car?

For comparison, it is not possible to drive an electric car with petrol and, on the other hand, to drive a car with a petrol engine using the battery of an electric car. That is more than a matter of course but will be more or less a mystery in other cases.

2.4 SUMMARY

In the study of nanosystems, three questions have to be considered carefully:

(1) What parameters are characteristic?
(2) What kind of physical laws have to be applied?
(3) What is the level of description?

We gave some typical examples for nanosystems: one from materials science; another from food chemistry; and a third from biology, from medicine in connection with brain research. These three examples are sufficient for explaining the most basic and important questions regarding nanoscience and nanotechnology.

Materials science (cluster states)

We considered a simple cluster consisting of aluminum atoms. It is made up of approximately 500 atoms, and it has a certain temperature. For this Al cluster, realistic model calculations have been performed. Such small systems behave nonharmonically already at relatively low temperatures, which can only be treated reliably within the frame of molecular dynamics. Here the most important point is the interaction between the atoms. In the case of our Al cluster, a realistic pair potential has been developed on the basis of the electronic structure of the atoms, and this has been used in the molecular dynamics calculations.

The cluster is obviously first in a metastable state and transforms spontaneously into a stable state. When a cluster transforms from the metastable to the stable state, there are various possibilities for that. In other words, there is a *bifurcation* in the sense of chaos theory. At the bifurcation point, nature plays dice in order to decide on which of the various branches the cluster will finally rest (Fig. 5). Without doubt, this is an interesting phenomenon.

Although the Al clusters represent *inorganic* systems, there are certain analogies with biological systems. There is obviously no definite line between materials science and biology — at the nanolevel, of course. This point is of particular relevance and indicates that already relatively simple systems exhibit complex behavior with respect to structure and dynamics. In particular, there are bifurcation phenomena in connection with such clusters.

The bifurcation phenomenon reflects the behavior of single clusters, and this specific effect comes into play by the mutual influence of temperature and the interaction between the atoms forming the cluster. There is a certain kind of independent creativity, and this creativity does not come from outside.

Nanosystems are important and interesting because they behave in relevant cases quite differently from systems used in micro- and macrotechnology. However, many researchers discuss them on the basis of traditional thinking. Why? They very often study the effects by means of static building blocks, as we do in connection with micro- and macrosystems. This view is, however, too narrow, because nanoclusters can be in an excited state (like atoms). After a certain time the excited cluster transforms spontaneously to the ground state without external influence. There can be more than one ground state and it is quite a matter of chance to what ground state the cluster transforms from the excited state.

What parameters are relevant in connection with nanosystems? In the description of nanosystems, three points are of considerable importance: the particle number N, the interaction potential between the particles, and the temperature of the system under investigation. Clearly, other specific parameters are important as well and determine the behavior of such systems.

Food chemistry

Regarding food chemistry, a lot of changes could occur. As a typical example we have mentioned the nanopizza.

The menu of an Italian restaurant often offers more than 20 pizza types: a pizza made of cheese, a pizza made of spinach, a pizza made of fish, and so on. Due to nanotechnology, this situation could be changed fundamentally. In the future, we will possibly have all these pizza types in one and, as we have pointed out, this is a question of temperature.

Brain research

The manipulation of atoms, molecules, etc. will allow us to construct new technological worlds and will bring fundamentally new possibilities also in the field of medicine, in particular within the field of brain research.

Just in the case of nanobiotechnology, big changes are expected already in the near future. It has been speculated that through nano-technology our bodies will be transformed into illness-free, undecaying systems of permanent health. In addition, it has been prognosticated that it will take about 30–50 years to develop nanotechnological means for the creation of superhuman intelligence.

First attempts have been made to increase the intelligence of animals by changing nanotechnologically certain brain functions. For example, a kind of genetic turboaccelerator has been developed for mousy intelligence. Molecular biologists, at Princeton added a gene to the brain of a mouse, raising the production of NR2B protein. The improved brains of the mice used this extra NR2B protein to enhance brain receptors, helping the animals to solve puzzles much faster. It is of particular interest that the human brain uses an almost identical protein. Thus, it is obvious to assume that we will learn to increase our own effective intelligence or to cure Alzheimer's disease (in the case of another treatment, of another input).

Do we really know what we do in such cases? When we change certain brain functions, what about the other brain functions that are responsible for other tasks? We have to analyze such situations *before* we perform such nanotechnological manipulations. For the estimation of the results, reliable theoretical (computational) methods and models have to be applied. The following question arises: Are the theoretical tools developed so far sufficient for the description of such biological phenomena? This question has to be investigated carefully.

However, are we really able to make *complete* statements about a human brain? Do we really know how it works? In other words, is a human able to understand his own brain? As we have outlined above, here Gödel's theorem comes into play, and it says that no mathematical system can make complete statements about itself. This theorem has to be taken into account because a mathematical scheme is needed for the theoretical analysis of a human brain!

In conclusion, when we transfer Gödel's theorem to the brain situation, we come to the result that a human can never completely understand his own brain. Such a self-related analysis is obviously not completely possible.

We know a lot about the human brain. The problem is that we do not know what we do *not* know about it. In other words, we have to be careful when we "play" with the human brain. Due to Gödel's theorem we cannot completely know what we do when we change the brain functions.

Relevant investigations have also been made within the framework of behavior research. From all these statements in connection with behavior research, we may conclude that the world view of human beings — the perception of the world in front of and around us — is dependent on the biological structure of man, i.e. on the brain functions. However, this world view is essential for individuals for survival, at least in the early phase of evolution, and each phase later is based on the biological structure developed in the earlier phase of evolution.

If we change any specific brain function nanotechnologically, we possibly also change the world view and possibly disturb the relationship between the human being and his environment, which can lead to a disaster, just as we have learned from the chick experiment within the framework of behavior research.

Even when the new world view — produced by nanotechnological changes of the brain functions — is more powerful and efficient than the old one, the human being is in general no longer adapted to the environment and is also no longer in accord with all the other human beings. To conclude, such a situation can have the effect that such a human being is not able to survive; we cannot explicitly exclude such a terrible scenario.

In conclusion, a human being is obviously not able to recognize what we have called "basic reality"; exactly the same should be true of other species, such as the turkey. Man uses certain information from basic reality for the construction of a species-dependent reality, which is tailor-made for him, and he recognizes this human-dependent reality on various levels — levels of reality.

We may state, quite generally, that changes of brain functions can lead to serious problems. All brain functions are more or less interrelated. If we change one brain function positively, other brain functions are influenced automatically and simultaneously.

Chapter Three

LEVELS OF THEORETICAL
DESCRIPTION

■ ■ ■

Nanosystems are exciting and behave in relevant cases quite differently from systems used in micro- and macrotechnology. Without doubt, nanotechnology and nanoscience, respectively, will have a large impact on the future life of human beings.

In Chap. 2, we firstly discussed a simple aluminum cluster. Although this cluster system is inorganic in character, it shows strong similarities to biological systems.

Secondly, we emphasized that food chemistry will possibly change our situation on the earth radically; we demonstrated that with the help of a nanopizza.

Thirdly, we underlined the relevance of nanotechnology to medicine, particularly brain research. Concerning brain functions, let us deal briefly with the most critical point.

3.1 PHYSICAL MODELING OF BRAIN FUNCTIONS

In the case of nanotechnological manipulations of brain functions, we have to precisely know what we do. This is possible only if reliable theoretical models are available. The following must be emphasized once more: when we want to describe the brain functions with the theoretical models of physics, the world view on which these theoretical models are based has to be consistent with the real mechanisms inside the brain. The world view inside the brain, which creates unconsciously the perceptions in front of and around us, comes into play through a projection: reality outside is projected onto the brain structures. In other words, the unconscious creation of the species-dependent reality is based on the "projection principle" (Chap. 2 and App. B, Sec. B.6).

As we know, physics presently works within the so-called "container principle," and here everything is embedded in space and time, i.e. space–time is considered as a container; in this case Fig. 14 is valid. In the case of the projection principle Fig. 15 is relevant, i.e. here we always have a transformed reality. Within projection theory the picture in front of us is a transformed reality; within the frame of the container principle it is believed that the container is the basic, objective reality. Physics is presently based on the container principle and it is assumed that there is a one-to-one correspondence between the picture and the reality outside (Fig. 14).

Clearly, the projection principle and the container principle are not consistent with each other. This can lead to serious problems when we try to manipulate nanotechnologically the brain functions on the basis of the "container world view."

3.2 PARAMETERS

What about the parameters characterizing the nanosystem? Two points are relevant:

(1) We have to recognize what *kind* of parameters are characteristic. This is not trivial!

(2) How sensitive are the properties of nanosystems with respect to variations of these parameters, with respect to variations of the characteristic functions? This is also not trivial and needs to be investigated carefully.

This situation dictates the method of description, and it also dictates the level of reality. Do we work at the macroscopic level, at the microscopic level, or at a level between these two limiting cases?

As we know, in nanoscience and nanotechnology one works at the microscopic level, at the atomic level, not only theoretically but also experimentally, and of course technologically too. This in particular means that the various disciplines grow together. This is an important point, and we will explain why.

In conclusion, in nanoscience and nanotechnology we are working at the atomic level, theoretically as well as experimentally. Here the experimental situation is relevant, also with respect to technical applications.

3.3 SCANNING TUNNELING MICROSCOPE

With the development of the scanning tunneling microscope, approximately 25 years ago, nanoscience (nanotechnology) became an important scientific and technological discipline, since for the first time single atoms could be moved in a controlled manner from one position to another, and we learned to manipulate matter at its ultimate level.

3.4 ULTIMATE LEVEL

But what does the term "ultimate level" mean? This is the *lowest* level at which the properties of our world emerge, at which functional matter can exist. This is particularly the level at which biological individuality comes into existence. Just the last point is of basic relevance.

This situation can be defined in absolute terms. In this connection Mark Rathner remarked[25]: "...through nanotechnology, we can make materials whose amazing properties can be defined in absolute terms: This is not only the strongest material ever made, this is the strongest material it will ever be possible to make."

This is not only the case for materials (particularly, nanostructured materials) but also biological structures such as DNA, enzymes, and proteins, which work at the nanoscale, building up, molecule by molecule, macroscopic biological systems, which we call trees, humans, and all the other things having their typical intimate features.

3.5 DESCRIPTIONS AT THE ULTIMATE LEVEL

In the description of phenomena at the ultimate level, adequate theoretical methods have to be used. This is the case when we design nanomachines, and also when we want to determine or construct the properties of new materials or to change biological systems nanotechnologically. In particular, the challenges in connection with nanomedicine require sophisticated theoretical descriptions. In order to be able to work also here at the ultimate level, completely new disciplines come into existence, for example computational neurogenetics. Let us briefly make some principal remarks.

3.5.1 Brain Functions and Computational Neurogenetics

The impact of nanotechnology will be tremendous. The manipulation of atoms, molecules, etc. will allow us to construct new technological worlds and will bring fundamental new possibilities in the field of medicine. We have already emphasized that in our discussions above and we will deepen this point a little bit.

Let us start with an example: it has been reported in the August 25 (2006) issue of *Science*[27] that scientists at the SUNY Downstate Medical Center have found a molecular mechanism that maintains memories in the brain, namely by persistent strengthening of synaptic connections between the neurons. The scientists were able to demonstrate that by inhibiting the molecule, long-term memories can be erased. Erasing the memory from the brain does not mean that the capability to relearn memory must be lost, i.e. this erasing process does not prevent that. The scientists could demonstrate these effects by an enzyme molecule with the name "protein kinase M zeta."

The detailed knowledge about such processes could be of fundamental importance for medicine, for example in connection with Alzheimer's disease. In general, such findings could in the future be helpful in treating chronic pain, posttraumatic stress disorder, memory loss, etc.

In the description of such phenomena, we must have recourse to the basic laws of theoretical physics. However, we cannot rule out the possibility that the usual laws are not sufficient, because the basic laws of theoretical physics have not yet been confronted with this kind of phenomena, for example the case we have just discussed — a molecular mechanism that maintains memories in the brain. It is one of the goals of computational and theoretical nanoscience to understand and to describe such molecular mechanisms. In particular, the prognosis is relevant in order to keep nanoscience under control.

In this connection it should be noted that within the frame of nanoscience and nanotechnology the relatively new discipline of computational neurogenetics will be of considerable importance, where the interaction between brain functions and genes is of specific interest[28]: "With the recent advancement of genetic research and the successful sequencing of the human and other genomes, more information is becoming available about the interaction between brain functions and genes, about genes related to brain diseases (e.g., epilepsy, mental retardation, etc.) and about gene-based treatment of them. It is well accepted now that brain functions are better understood and treated if information from the molecular and neuronal level is integrated ... For this purpose, computational models that contain genetic and neuronal information are needed for modeling and prognosis."

Such a scientific program (modeling and prognosis) is necessary in order to eliminate or to reduce the threats that appear in connection with the manipulation of brain functions. Just the prediction of the evolution of certain artificial biological structures is relevant. However, here also we must always keep in mind that there are possibly principal limits to the complete description of brain functions. We have discussed this point by means of Gödel's theorem and on the basis of the "principle of level analysis" in Chap. 2, Sec. 2.3.2. In the following we will discuss phenomena in nanoscience where brain functions and mental states are not involved.

Chapter Four

SOME REMARKS ON NANOENGINEERING

■ ■ ■

Nanoengineering and nanodesign will certainly be among the most important future disciplines within the field of nanotechnology. In this monograph we will not give an overview of these developments, which have been discussed in the literature. Instead, let us present some typical features on the basis of an example — an electrical nanogenerator.

4.1 NANOGENERATOR DESIGN

It is well known that nanosystems are very small. The following example demonstrates that impressively. An electrical nanogenerator has been modeled; the details are given in Ref. 26. In principle, it could be produced atom by atom, but this kind of production would not be very efficient. Nevertheless, let us briefly discuss this computer-designed machine; the structural and dynamical behavior of the generator has been studied through a realistic molecular dynamics calculation.

This nanogenerator is shown in Fig. 20. It has a size of only about 20 nm. One can recognize that this nanomachine is small but relatively complex. It consists of five parts: a winding (made of aluminum) with two axes, an isolating kernel (krypton) which stabilizes the winding, two bearings (made of silicon), and a substrate on which the generator is positioned. The assembled nanogenerator can rotate with more

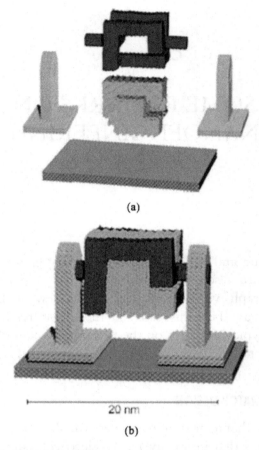

(a)

20 nm

(b)

Fig. 20   (a) The nanogenerator for electrical power consists of five parts: a winding with axles, an isolating kernel, two bearings and a substrate. The winding consists of 25,433 aluminum atoms, while the rotor kernel has been assembled from 11,341 krypton atoms. The bearings are treated as rigid objects made of silicon. (b) The assembled nanogenerator is able to rotate with 5×10^9 revolutions per second. (© 2006, American Scientific Publishers.)

than 10^9 revolutions per second. For revolutions greater than 10^9, the system ruptures.

The number 10^9 for the revolutions per second is a very large. If the wheels of a car could rotate with such a large number of revolutions, the car would circle the earth more than 80 times in a second. This is unimaginable.

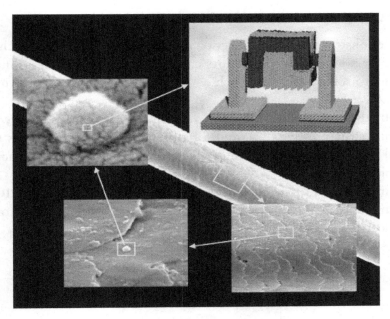

Fig. 21　This figure shows the size of the electrical nanogenerator in comparison with a hair. The size of the generator is approximately 20 nm, and the hair has a diameter of 80,000 nm; that is to say, the diameter of the hair is 4000 times larger than the size of the generator. (© 2006, American Scientific Publishers.)

Figure 21 shows the size of the nanogenerator in comparison with a hair. The straight line represents the hair. The nanogenerator could be positioned on a very small area of the hair. The hair has a diameter of 80,000 nm; in other words, the diameter of the hair is 4000 times larger than the size of the nanogenerator, which has an extension of 20 nm. This is also almost unimaginable.

Another example: approximately one million such electrical nanogenerators could be arranged side by side within a space interval of 1 cm.

In conclusion, the space–time behavior of nanosystems is quite different from those we experience in everyday life. Clearly, this situation dictates the methods of description in nanotechnology, which are different from those we use in the construction of macroscopic systems (cars, houses, and so on).

4.2 CONSTRUCTION AND PRODUCTION OF NANOSYSTEMS

In the construction and production of macroscopic machines (such as a car or a locomotive), the parts can be treated as continuous units without atomic structure; such a "smooth" workpiece is shown in Fig. 22.

In the nanoregion, we have very small granular systems with atomic structure (Fig. 23), and for the shape and the dynamics the total number (N) of atoms (molecules, etc.) is relevant; in the macroscopic case, not.

In the case of macroscopic systems we have approximately 10^{23} atoms, and in the case of *nanosystems* we have only 10^2 atoms (molecules) or something like that.

Fig. 22   A workpiece as it is used in macrotechnology, for example in the production of a car. The parts can be treated as continuous units without atomic structure. The form of such workpieces is in a good approximation independent of temperature. (© 2006, American Scientific Publishers.)

Fig. 23   In the nanoregion, the systems under investigation have to be treated as granular units with atomic structure. (© 2006, American Scientific Publishers.)

More relevant is the relative number (n) of atoms at the surface: $n = N'/N$, where N' is the number of atoms at the surface (see also Chap. 2.1.3).

In the case of macroscopic systems, we have for n the value of 10^{-8}; there are practically no surface atoms, but only bulk states. For nanosystems we have $n = 0.5–1$. In the case of $n = 1$, all atoms would behave like surface atoms. Because of this big difference concerning n, nanosystems behave quite differently from macroscopic systems (which are in the bulk state, with almost no surface atoms). One may state that the properties are not comparable with each other, even in the case of the same type of atoms or molecules. For example, for aluminum we have the following behavior:

In the nanocase, we have no spontaneous transformations as we observe in the case of nanoclusters (see Chap. 2). Furthermore, the melting temperature of aluminum in the bulk is 933 K; the melting temperature of the Al nanoclusters studied in Chap. 2 is only 290 K. This is in fact a big difference.

Why is the number of surface atoms the essential factor? Answer: Because the atoms at the surface behave quite differently from those in the bulk that are positioned in the inner part of the system (workpiece).

4.2.1 Nanosystems

In contrast to the bulk states, we have strong anharmonic effects at the surface, and the positions of the atoms (or molecules) are no longer well defined, in contrast to macroscopic systems. This effect increases with increasing temperature.

The system does not remain in the zero temperature state [Fig. 24(a)], but we obtain, with increasing temperature, a strongly disturbed structure; a typical structure is shown in Fig. 24(b), which is based on a realistic molecular dynamics calculation. In contrast to macroscopic systems, the whole shape of the nanosystem is disturbed, and this is due to the unavoidable temperature.

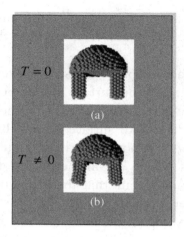

Fig. 24   (a) Nanosystem at zero temperature ($T = 0$). (b) The same nanosystem for $T \neq 0$; the structure is distinctly disturbed. (© 2006, American Scientific Publishers.)

4.2.2 Macroscopic Systems

If we work below the melting temperature, the shape of the parts of a macroscopic machine does not vary with temperature; clearly, we have a certain thermal expansion, but this is a uniform effect and does not affect the shape of the workpiece.

Furthermore, the shape of a certain workpiece is hardly or not influenced by the other workpieces around it. This is a relevant point in connection with the production of macroscopic machines. In other words, the shape of the isolated workpiece is exactly the same as in the composed state of the machine. So, we may produce the parts of a macroscopic machine independently of the other parts that form the machine.

In conclusion, the parts of a macroscopic machine can be treated as continuous blocks without atomic structure, and their shapes do not vary with temperature and are also not influenced by the other parts of the machine. All this is trivial and is a matter of course. The constructions of everyday life are based on this feature.

The bulk states are equivalent to the situation which we have discussed in connection with macroscopic machines, which have no atomic structure. Here, in the case of the bulk states, the number of surface atoms is relatively small, and we have the situation as in the case of macroscopic machines.

Again, the shape of certain parts of a macroscopic machine does not vary with temperature; clearly, we observe here the effect of thermal expansion, but this is a uniform effect and hardly influences the shape of the workpiece.

In summary, the bulk states of a nanosystem are influenced by temperature too, but here also only the thermal expansion is relevant.

4.2.3 The Influence of other Workpieces

Furthermore, in the case of macroscopic machines the shape of a certain workpiece is not influenced by the other workpieces around it, as we have outlined above. Therefore, all parts of a macroscopic machine can be produced separately, and we do not have to consider the others when we produce a specific workpiece.

If the macroscopic machine has n parts, we produce each of these parts independently of the others. After that we can put these n parts together in order to get the machine.

In the case of nanosystems, for example a nanomachine, the situation is quite different. Why? Let us briefly discuss this point and let us consider a typical example.

A certain part of the nanosystem, which is surrounded by the other parts of the nanomachine, might have the shape given in Fig. 25(a). The other parts of the system are symbolized in the figure by an abstract configuration [the frame in Fig. 25(a)].

However, if the same system is isolated we obtain another form for this part of the nanosystem [see Fig. 25(b)]. The two pieces have the same temperature, T_a.

In conclusion, the isolated part is different from that which is surrounded by the other parts of the nanomachine. In other words, this system cannot be used for the nanomachine, because it is not identical with that inside the nanomachine.

This must have consequences. If the engine of a car is defective, we can exchange it for a new engine. However, such a procedure does not work at the nanolevel. This is because the shape of a part inside

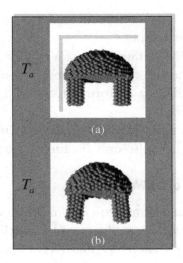

Fig. 25   (a) Part of a complex nanosystem inside the machine. (b) The same part outside the machine. The temperature of both systems (inside and outside the nanomachine) is T_a.

the nanomachine is different from that outside the machine, which is not surrounded by other parts. As demonstrated in Fig. 25, this is due to the unavoidable temperature and surface effects.

4.3   WHAT DIRECTIONS DO WE EXPECT IN NANOTECHNOLOGY?

What directions are expected in nanoscience and nanotechnology? In the reduction of a machine from the meter to the nanometer realm, the atomic structure becomes relevant, as we have seen in the case of the electrical nanogenerator (Fig. 21).

Nevertheless, this electrical nanogenerator works like a macroscopic machine we know from everyday life. In other words, the machine is a strongly reduced system, but it is still macroscopic in character. Why can we draw this conclusion? Let us discuss this relevant point.

At the nanolevel, new properties emerge that are not known at the macroscopic level. We have recognized in the case of the aluminum nanocluster (Chap. 2) that inorganic nanosystems can have biological features, and this kind of nanoproperty does not appear in the case of the electrical nanogenerator. It is small, very small, but behaves more or less like a macroscopic machine, such as a car or a locomotive.

Instead, nanosystems (or nanomachines), which are produced by self-organizing processes (self-assembly), will be at the center of future nanoscience and nanotechnology. Such self-organizing processes are in principle able to produce the typical effects at the nanoscale. Therefore, in the future, self-organizing processes will belong to the heart of nanoscience and nanotechnology.

Again, such nanomachines (like the electrical nanogenerator) are strongly reduced systems, but are still macroscopic in character, and new properties, which emerge at the nanolevel, do not appear here. Such machines would be produced "atom by atom" with the specific operations by the mechanic, using, for example, the famous scanning tunneling microscope.

In the case of self-organizing processes, the systems develop according to the physical laws but without the specific operations by the engineer. In the case of the nanopizza (Chap. 2, Sec. 2.2), there is no cook who puts the spinach (or the fish) onto the pizza dough, but the spinach (or the fish) emerges through a self-organizing process without the specific operations by the cook.

This is a situation quite in analogy with that we observe in biology. In nanoscience (nanotechnology) we deal with open, self-organizing systems that communicate with their surroundings. Let us now make more statements about nanoengineering, and after that about self-organizing systems.

4.4 NANOENGINEERING

The future nanoengineer has to prepare the intitial conditions, i.e. the initial system that develops by self-organizing processes in the course of time. However, the properties of the new systems will also be dependent on the structure of the physical laws, which describe the nanoprocess as a function of time. This self-organizing process transforms the initial system into the new system.

4.4.1 What Theoretical Tools?

What tools (theoretical tools) does the nanoengineer need for the construction, for the design, of the nanomachine or any other nanosystem? The selection of adequate tools is one of the most relevant tasks in theoretical and computational nanoscience and nanotechnology.

In nanoscience (nanotechnology) we work at the ultimate level, where the properties of materials, biological systems, etc. emerge (see in particular Chap. 3, Sec. 3.3), and therefore we have to apply the basic physical laws. Do we really know these basic physical laws? When we talk in the following of basic laws, we always mean the best laws that are presently known.

This in particular means that we have one theory for all phenomena. At the nanolevel we have one theory for all phenomena, and this is given by the laws of theoretical physics. We need those basic laws of theoretical physics, which are formulated for the nanolevel. We do not need the theoretical structures of elementary particle physics, string theory, or cosmology.

4.4.2 Remark

In traditional technologies (micro- and macroengineering), engineers normally do not work at the ultimate level. They use more or less phenomenological descriptions, which in general cannot be deduced from the basic physical laws; each (traditional) technological discipline has its own description.

4.4.3 Theoretical Modules

Clearly, for each discipline (nanoelectronics, nanomedicine, etc.) we will have a specific theoretical module, i.e. a specific theoretical architecture, but each module will be developed by means of the laws of theoretical physics without leaving this level (see also Fig. 26).

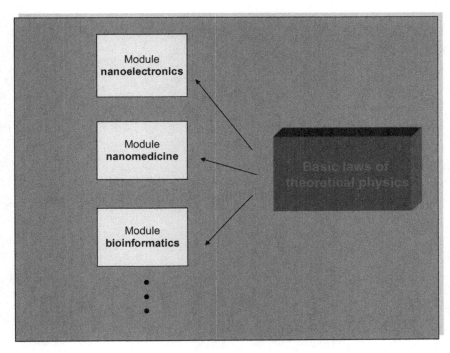

Fig. 26   We have various disciplines. For these disciplines, we will have specific theoretical modules (architectures), developed by the laws of theoretical physics, without leaving the level of theoretical physics.

Chapter Five

SELF-ORGANIZING PROCESSES

■ ■ ■

We have learned from the last chapter that the future nanoengineer will work at the level of theoretical physics; in other words, he will need the laws of theoretical physics. Furthermore, we remarked that self-organizing processes are essential for the design of new nanosystems. In this chapter, we will make some principal statements about self-organizing processes.

Let us repeat once more that in the case of self-organizing processes the systems develop according to the basic physical laws, but without the specific operations by the mechanic. In the development of macroscopic systems, self-organizing processes are not the essential factor.

5.1 THE PHYSICS OF BECOMING

The nanosystems develop in the course of time from a certain initial state to a final state:

$$\text{initial state} \rightarrow \text{final state.}$$

Because of the time dependence we go from the physics of being to the physics of becoming.

In conclusion, self-organizing processes belong to the physics of becoming. The scientists have a lot of experience with the physics of being (with stationary systems). All the specific structures of the atoms, molecules, solids, etc. are well described by the physics of being, i.e. by stationary equations. But we have little experience with the physics of becoming, which is relevant to self-organizing processes, described by nonstationary equations.

5.2 ON THE DESCRIPTION OF SELF-ORGANIZING PROCESSES

In Ref. 26, a new method has been developed, which could be relevant to the description of self-organizing processes. It is the so-called potential morphing method (PMM).

In the PMM we have a certain initial state, A, which has to be known, and there is a certain final state, B, that is unknown and must be calculated and predicted, respectively. The PMM allows us to describe the transition from A to B as a function of time τ:

$$A \xrightarrow{\tau} B. \tag{1}$$

Transition (1) has to be investigated when we try to analyze self-organizing processes.

5.2.1 Application of Schrödinger's Equation

In most cases, nanosystems behave quantum-mechanically and we have to apply Schrödinger's equation.

Remark

For nanosystems the molecular dynamics method is of particular relevance, but within the frame of such calculations we need the interaction potential between the atoms (molecules, etc.), and these interactions can only be understood and determined on the basis of quantum theory.

Schrödinger's equation for the stationary case is given by the well-known relation

$$i\hbar\frac{\partial\Psi(\mathbf{r},\tau)}{\partial\tau} = -\frac{\hbar^2}{2m_0}\Delta\Psi(\mathbf{r},\tau) + V(\mathbf{r})\Psi(\mathbf{r},\tau), \qquad (2)$$

and for the nonstationary case we have

$$i\hbar\frac{\partial\Psi(\mathbf{r},\tau)}{\partial\tau} = -\frac{\hbar^2}{2m_0}\Delta\Psi(\mathbf{r},\tau) + V(\mathbf{r},\tau)\Psi(\mathbf{r},\tau), \qquad (3)$$

where $\hbar$ is Planck's constant, $\Psi(\mathbf{r},\tau)$ is the wave function, and V in both equations is the interaction potential. $\mathbf{r}$ summarizes the positions of all particles of the nanosystem: $\mathbf{r} = (\mathbf{r}_1, \mathbf{r}_2, \ldots, \mathbf{r}_N)$. The term with m_0 is the short form for all particles involved in the process $(m_0 : m_1, m_2, \ldots, m_N)$.

"Stationary" means that the potential V is not dependent on time τ and we get Eq. (2); if V is dependent on τ, we deal with the nonstationary case and Eq. (3) is valid.

The PMM is based on these equations — on the stationary and the nonstationary wave equation. In the transition $A \overset{\tau}{\to} B$ [Eq. (1)], the initial state A and the final state B are expressed by these *stationary* equations [see in particular Eq. (2)]:

$$i\hbar\frac{\partial\Psi_A(\mathbf{r},\tau)}{\partial\tau} = -\frac{\hbar^2}{2m_0}\Delta\Psi_A(\mathbf{r},\tau) + V_A(\mathbf{r})\Psi_A(\mathbf{r},\tau), \qquad (4)$$

$$i\hbar\frac{\partial\Psi_B(\mathbf{r},\tau)}{\partial\tau} = -\frac{\hbar^2}{2m_0}\Delta\Psi_B(\mathbf{r},\tau) + V_B(\mathbf{r})\Psi_B(\mathbf{r},\tau). \qquad (5)$$

On the other hand, for the transition from A to B Schrödinger's nonstationary equation holds [see Eq. (3)]:

$$i\hbar\frac{\partial\Psi_{AB}(\mathbf{r},\tau)}{\partial\tau} = -\frac{\hbar^2}{2m_0}\Delta\Psi_{AB}(\mathbf{r},\tau) + V_{AB}(\mathbf{r},\tau)\Psi_{AB}(\mathbf{r},\tau). \qquad (6)$$

Here the potential is dependent on time, but not in connection with Eqs. (4) and (5).

5.3 TWO CASES

Now, we can distinguish between two cases. We start with case 1 and define the following:

Case 1

In the transition from A to B, $A \xrightarrow{\tau} B$ [Eq. (1)], the initial system A is known, and we have system B in mind. Then, we try to realize system B, theoretically as well as experimentally.

How can we realize what we have in mind? In other words, how can we realize system B? The answer is given as follows: Schrödinger's equation

$$i\hbar \frac{\partial \Psi_{AB}(\mathbf{r}, \tau)}{\partial \tau} = -\frac{\hbar^2}{2m_0} \Delta \Psi_{AB}(\mathbf{r}, \tau) + V_{AB}(\mathbf{r}, \tau)\Psi_{AB}(\mathbf{r}, \tau)$$

[Eq. (6)] holds for the transition from A to B: $A \xrightarrow{\tau} B$. We vary the potential $V_{AB}(\mathbf{r}, \tau)$ until we have the system of our choice, i.e. until we have system B. In other words we produce system B using system A as the starting point and vary the environment (the interaction of the system with the environment).

In case 1, we have to construct an environment in such a way that the self-organizing process leads to the system of our choice; we vary the environment until we have a system with such properties that we want to have.

Case 2

Case 2 is of particular relevance: we prepare the initial system A and put it into a given environment. Then the experimentalist does not know the final state B at the beginning of the self-organizing process. In other words, the new system, the outcome, remains unknown until the self-organizing process is finished.

This situation must be reflected in the interaction potential and we come to a symbol for V with a question mark: we have $V_{A?}(\mathbf{r}, \tau)$, instead of V with B, i.e. $V_{AB}(\mathbf{r}, \tau)$.

In other words, the letter B is simply replaced by a question mark, and this is because we do not know the final state B at the beginning of the self-organizing process. Instead of this transition,

$$A \xrightarrow{\tau} B$$

(case 1), we have

$$A \xrightarrow{\tau} ? \tag{7}$$

This equation defines case 2. Instead of this equation,

$$i\hbar \frac{\partial \Psi_{AB}(\mathbf{r}, \tau)}{\partial \tau} = -\frac{\hbar^2}{2m_0} \Delta \Psi_{AB}(\mathbf{r}, \tau) + V_{AB}(\mathbf{r}, \tau)\Psi_{AB}(\mathbf{r}, \tau)$$

[Eq. (6), case 1], we get

$$i\hbar \frac{\partial \Psi_{AB}(\mathbf{r}, \tau)}{\partial \tau} = -\frac{\hbar^2}{2m_0} \Delta \Psi_{AB}(\mathbf{r}, \tau) + V_{AB}(\mathbf{r}, \tau)\Psi_{AB}(\mathbf{r}, \tau), \tag{8}$$

and this equation is characteristic of case 2. In connection with case 1, we know the new system already at the beginning. Regarding case 2, we do not know the new system at the beginning. This point has to be discussed in more detail.

5.4 DANGEROUS SITUATIONS

Case 2 (the transition $A \xrightarrow{\tau} ?$) can lead to dangerous situations. Let us give an example, discussed in the literature. Such a final state, $\Psi_{A?}(\mathbf{r}, t)$, as a result of Eq. (8), could be, for instance, a self-replicating device, i.e. a self-replicating nanodevice, a certain kind of nanorobot. Such a nanorobot could get out of control and could spread unrestrainedly on the earth, having the effect that everything on the earth transforms into a differentiationless mass. In other words, life on the earth would be destroyed.

Without doubt, this is a horror scenario! We have not invented this example, but the self-replicating nanorobot has been discussed seriously in the literature.

Is such a horror scenario exaggerated? The probability for such and similar situations is small — probably very small. But it is not exactly zero. We do not always know with certainty what kind of system the self-organizing process emerges.

On the other hand, due to nanotechnology there will be a lot of positive facts. Cancers will be cured, along with most other ills of the flesh. Aging, or even routine death itself, might become a thing of the past.

Nevertheless, the threats are immense, as we have just discussed in connection with self-replicating nanorobots. The potential threats in nanotechnology are large — probably much larger than those in the field of nuclear energy.

5.5 WHERE ARE THE LIMITS?

In nanoscience (nanotechnology) we move between two limiting cases: infinite life and total destruction. This is perhaps somewhat exaggerated, but all these things are conceivable. The reason is obvious: we are able to work at the ultimate level; at this level biological individuality comes into existence and, therefore, we can make very fundamental changes.

5.6 RELIABLE MODEL CALCULATIONS ARE NECESSARY

We have to keep nanotechnology under control. In particular, we have to keep the self-organizing processes under control. We must know what we do and, thus, before we produce nanosystems by self-organizing processes we must perform reliable model calculations.

We must understand the self-organizing process and should be able to estimate the product of the self-organizing process (the final state, the outcome, the new system). That is to say, the final state $\Psi_{A?}(\mathbf{r}, t)$ must be known, before we start to produce the nanosystem. In other words, what is the question mark in connection with $\Psi_{A?}(\mathbf{r}, t)$? This exactly has to be known before we start the self-organizing process in the laboratory.

Chapter Six

THE BASIC LAWS OF PHYSICS

■ ■ ■

In this chapter, we will make some remarks on the physical laws that
we use in nanoscience and nanotechnology. We will concentrate on
self-organizing processes, which will undoubtedly play an important
role in the development of nanosystems in physics, chemistry, and
biology. But changes of already existing systems are also concerned.
In all these cases the laws of theoretical physics are important.

We have studied self-organizing processes in the last chapter. Here
the transition of state A to state B in the course of time τ has been
investigated. We may state quite generally that the final state B of such
self-organizing processes will be dependent on the initial state A, how
it has been prepared at the beginning. From A the system develops in
the course of time τ toward B: $A \xrightarrow{\tau} B$.

But there is another relevant point: the final state B will also be
dependent on the structure of the physical laws, which describe the
transition from A to B as a function of time τ.

6.1 STRUCTURE OF THE BASIC LAWS

Since we work in nanoscience (nanotechnology) at the ultimate level,
the structure of the "basic" physical laws is of particular importance.
But we should not have the illusion that we will ever have the final
laws of basic physics in our hands — our hands are not so large.
Human beings are obviously not be able to recognize what is called

the "absolute truth". In Chap. 2, we have discussed this point in connection with the principles of evolution.

So far we have used Schrödinger's equation, and this equation is presently considered as a "basic" physical law since it could not be deduced from other physical laws.

Question: Is the transition $A \xrightarrow{\tau} B$ unquestionably described by $i\hbar\partial\Psi_{A?}(\mathbf{r},\tau)/\partial\tau = -\hbar^2(2m_0)\Delta\Psi_{A?}(\mathbf{r},\tau) + V_{A?}(\mathbf{r},\tau)\Psi_{A?}(\mathbf{r},\tau)$, which is, as we know, Schrödinger's equation [Eq. (8)]? Or is the connection between the wave function $\Psi_{A?}(\mathbf{r},\tau)$ and the interaction potential $V_{A?}(\mathbf{r},\tau)$ described in another way? In other words, is it described by a formula different from Schrödinger's equation?

Such and similar questions have to be investigated carefully in the future, since just in nanotechnology the laws of description have to be sufficiently reliable. However, when we change the structure of a basic physical law (for example Schrödinger's equation), we have to precisely explain why. We have to justify that there is a need for such modifications.

Are changes in connection with Schrödinger's equation really advisable? Yes, they are! Here we touch a very fundamental problem. Here *nanotechnology* touches a very fundamental problem. Let us see why.

6.2 WE HAVE A PROBLEM WITH TIME

The most important quantity in connection with self-organizing processes like $A \xrightarrow{\tau} B$ is time τ, and this is because the process from A to B develops in the course of time τ and, as we stated above, these processes are described by Schrödinger's equation. However, within Schrödinger's theory we have a problem with time τ! What kind of problem is it?

Our basic equation that describes the self-organizing process is given by Eq. (8):

$$i\hbar\partial\Psi_{A?}(\mathbf{r},\tau)/\partial\tau = -\hbar^2/(2m_0)\Delta\Psi_{A?}(\mathbf{r},\tau) + V_{A?}(\mathbf{r},\tau)\Psi_{A?}(\mathbf{r},\tau).$$

Question: What can we say about time τ, which appears in this formula? Answer: Time τ is still a classical parameter; the time behaves classically within the framework of Eq. (8) and this is a deficiency.

That is a quite general feature in connection with each form of Schrödinger's equation. Without doubt, Eq. (8) is a quantum-mechanical equation, but time is here still a simple classical parameter, as in Newton's mechanics. The character of time is not changed when we go from classical mechanics to quantum theory. This could lead to problems when we try to describe self-organizing processes quantum-theoretically, because the time is by definition an essential quantity in connection with self-organizing processes; the system develops in the course of time.

That time remains a classical quantity in the quantum-theoretical description has to be considered as a deficiency. The effect of this deficiency could be estimated if we had a quantum-theoretical picture for time. There are exciting developments in this field, absolutely exciting, but we will not discuss specific details in this monograph.

Within the conventional quantum theory we have operators for all variables, but not for the time. This is of course an accident. Within all versions of the usual quantum theory we are missing a quantum-mechanical aspect of time, i.e. an operator for the time!

Therefore, in introducing a quantum-mechanical aspect of time we have to assume that the structure of the physical laws will be changed basically.

6.3 SCHRÖDINGER, PAULI, PRIGOGINE

It was already argued by Erwin Schrödinger (1887–1961) and Wolfgang Pauli (1900–1958) that we need a quantum-mechanical aspect of time. But Ilya Prigogine (1917–2003) worked in this field too; he also strongly emphasized that we need a quantum picture for time.

All three of them argued that such a quantum aspect of time should be reflected by an operator description of time. In other words, time τ should no longer be a simple external parameter as in classical mechanics and in conventional (traditional) quantum theory.

For this purpose, Schrödinger introduced an operator for the time by means of this commutation rule: $[\widehat{T}, \widehat{H}] = i\hbar\widehat{I}$. Here $\widehat{H}$ is the Hamiltonian, $\widehat{I}$ the unit operator, and $\widehat{T}$ the operator for the time which has to be found.

Does there exist such a time operator $\widehat{T}$ within the conventional quantum theory? The answer is no! The operator $\widehat{T}$ is definitely not definable within the conventional quantum theory. In conclusion, the introduction of an operator for the time on the basis of $[\widehat{T}, \widehat{H}] = i\hbar\widehat{I}$ is not possible within the usual quantum theory.

Without going into details, we may state that a quantum-mechanical aspect of time (an operator for the time) cannot be introduced within the conventional quantum theory without changing the structure of the physical law itself, i.e. without changing the structure of Schrödinger's equation itself.

6.4 A QUANTUM-THEORETICAL ASPECT OF TIME IS NEEDED

However, as has been remarked above several times, in the description of self-organizing processes time τ is of particular relevance. As we stated above, the final state B of the process $A \overset{\tau}{\to} B$ will be dependent not only on the initial state A, but also on the structure of the physical laws which describe the transition from A to B in the course of time τ.

We may conclude from the analysis done by Schrödinger, Pauli, and Prigogine that the structure of the physical laws is changed when the nature of time is changed. In other words, when we introduce a quantum-mechanical aspect of time we have to assume that the structure of the basic physical laws will be changed simultaneously. Such a step is obviously unavoidable, but we need such a quantum aspect of time!

Just such laws would be changed that are responsible for the self-organizing processes in nanoscience. But it is obviously necessary to take such a step, because a reliable description of self-organizing processes demands in general a quantum time. As has been remarked, the effect due to a quantum time cannot presently be estimated. On the other hand, we should not have the illusion that we can ever formulate the *final* laws of physics. No doubt we will improve the basic laws, but we will (probably) never find their final form.

Chapter Seven

SUMMARY AND FINAL REMARKS

■ ■ ■

Let us briefly summarize the main results: in nanoscience and nanotechnology we work at the microscopic level, i.e. at the atomic level, not only theoretically but also experimentally, and of course technologically. This in particular means that the various disciplines (the various fields) grow together.

Nanosystems are interesting and behave in relevant cases quite differently from systems used in micro- and macrotechnology. However, many researchers discuss nanosystems on the basis of traditional thinking. Why? They very often study the effects by means of static building blocks, as we do in connection with micro- and macrosystems.

7.1 SYSTEMS AND PHENOMENA

7.1.1 Nanoclusters

As we have demonstrated in the case of nanoclusters, "traditional thinking" is obviously too narrow because nanoclusters can be in an excited state (like atoms). After a certain time the excited cluster transforms spontaneously to the ground state without external influence. There can be more than one ground state and it is quite a matter of chance to what ground state the cluster transforms from the excited state.

Although the aluminum clusters represent *inorganic* systems, there are certain analogies with biological systems. There is obviously no definite line between materials science and biology — at the nanolevel, of course. This point is of particular relevance and indicates that already relatively simple systems exhibit complex behavior with respect to the structure and dynamics. In particular, there are bifurcation phenomena in connection with such clusters.

The bifurcation phenomenon reflects the behavior of single clusters, and this specific effect comes into play by the mutual influence of temperature and the interaction between the atoms forming the cluster. There is a certain kind of independent creativity, and this creativity does not come from outside.

7.1.2 Materials Science

With the development of the scanning tunneling microscope, approximately 25 years ago, we learned to manipulate matter at its ultimate level. Single atoms could be moved for the first time in a controlled manner from one position to another, and nanoscience (nanotechnology) became an important scientific and technological discipline.

What does the term "ultimate level" mean? This is the *lowest* level at which the properties of our world emerge, at which functional matter can exist, at which biological individuality comes into existence.

This situation can be defined in absolute terms: *This is not only the strongest material ever made, this is the strongest material it will ever be possible to make.*[25]

Since we work at the ultimate level, we have to apply the basic laws of theoretical physics. This is an important point and has to be considered strictly.

7.1.3 Nanomachines

We discussed an electrical nanogenerator. This machine is very small, but it is still macroscopic in character and new properties, which normally emerge at the nanolevel, do not appear here, but such new properties appear within the frame of self-organizing processes.

Could we produce such an electrical nanogenerator at all? Yes, in principle we could, but only in principle, because such an "atom by atom" production (using, for example, the famous scanning tunneling microscope) would take a lot of time — probably too much time.

7.1.4 Self-Organizing Processes

Instead, self-organizing processes will be at the center; self-organizing processes will belong to the heart of future nanoscience and nanotechnology. Clearly, here also we work at the ultimate level and, therefore, we have to apply here also the basic laws of theoretical physics.

Self-organizing processes belong to the physics of becoming; physicists have a lot of experience with the physics of being (the structure of all atoms, molecules, and solids is well described by the physics of being), but we have little experience with the physics of becoming.

We have pointed out that the final state B of the process $A \overset{\tau}{\longrightarrow} B$ will be dependent on the initial state A, i.e. the final state will be dependent on the preparation of the system at the beginning. Furthermore, the final state B will be dependent on the structure of the physical laws, which describe the transition from A to B as a function of time τ.

Self-organizing processes can lead to dangerous situations, as we emphasized in Chap. 5, Sec. 5.4. When we put a certain system into a given environment, we do not know the outcome (the new system) at the beginning of the self-organizing process. This can lead to critical situations and, therefore, we have to analyze the self-organizing process theoretically before we start the production of a new system. Thus, reliable model calculations are necessary before we start a self-organizing process!

We have to keep nanotechnology under control. In particular, we have to keep the self-organizing processes under control. We need to know what we do and, thus, before we produce nanosystems by self-organizing processes we must perform reliable model calculations.

We must understand the self-organizing process and should be able to estimate the product of the self-organizing process (the final state, the outcome, the new system). In other words, the final state $\Psi_{AB}(\mathbf{r}, \tau)$ must be known, before we start to produce the nanosystem. In other words, what is the question mark in connection with $\Psi_{A\,?}(\mathbf{r}, \tau)$. This exactly has to be known before we start the self-organizing process in the laboratory. Otherwise we are possibly confronted with an uncontrolled situation.

Again, the transition $A \xrightarrow{\tau} B$ can lead to dangerous situations. In Chap. 5, Sec. 5.4 we have given an example. Such a final state B could be a self-replicating device, a certain kind of nanorobot. Such a nanorobot could get out of control and could spread unrestrainedly on the earth, having the effect that everything on the earth transforms into a differentiationless mass. In other words, life on the earth would be destroyed. Without doubt, this is a horror scenario!

Is such a horror scenario exaggerated? The probability for such and similar situations is small — probably very small. However, it is not exactly zero. We do not always know with certainty what kind of system the self-organizing process emerges as.

On the other hand, due to nanotechnology, there will be a lot of positive facts. Cancers will be cured, along with most other ills of the flesh. Aging, or even routine death itself, might become a thing of the past.

Nevertheless, the threats are large, as we have just discussed in connection with self-replicating nanorobots. The potential threats in nanotechnology are large — probably much larger than those in the field of nuclear energy.

We may state that in nanoscience (nanotechnology) we move between two limiting cases: infinite life and total destruction. The reason is obvious: we are able to work at the ultimate level; at this level biological individuality comes into existence and, therefore, we can make very fundamental changes that touch the possibilities "infinite life" and "total destruction."

In conclusion, reliable model calculations are necessary before we start a self-organizing process in the laboratory, in order to keep nanotechnology under control. This is an important point and may not be underestimated.

7.2 THEORETICAL DESCRIPTION

For a reliable description of nanosystems, particularly self-organizing processes, we need realistic models that have to be based on realistic physical laws, i.e. we have to choose an adequate level of description. As remarked in Chap. 6, this level is given by the basic laws of theoretical physics, and this is because we work at the ultimate level in nanoscience and nanotechnology.

In most cases nanosystems behave quantum-mechanically, and we have to apply Schrödinger's equation (see in particular Chaps. 5 and 6). However, within the framework of this equation time τ is still a classical parameter, already introduced by Isaac Newton — and this is of course a deficiency!

Since the time is an essential factor, in the description of self-organizing processes in nanoscience, we need a realistic quantum-theoretical picture for time, as already emphasized and investigated by Schrödinger, Pauli, and Prigogine. There are exciting developments in this field, but so far a quantum time has not been established in theoretical physics.

7.3 METHODS AND WORLD VIEW

The theoretical and computational methods and laws are dependent on the world view. For the deduction of a theoretical physical law we need a certain world view on which we can base our construction, which leads to the scientific law.

The present physical laws are based on the "container principle" (see App. B); within the frame of this principle everything (material bodies, etc.) is embedded in the space–time. However, we have to be careful, because there are relevant experiments indicating convincingly that the "projection principle" is valid and not the container principle. Within the projection principle is what we call reality not embedded in space and time, as is the case of the container principle, but reality is here projected onto space and time (see Chaps. 2 and 3, and App. B).

Presently, nanoscience is merely based on the traditional theoretical laws, which conform to the container principle, i.e. nanoscience is still dependent on the theoretical laws of traditional physics. The main methods (molecular dynamics, nonequilibrium molecular dynamics, quantum molecular dynamics, the Monte Carlo method, and multiscale modeling) have been quoted and briefly described in Chap. 2, Secs. 2.1.11 and 2.1.12.

7.3.1 Some Critical Remarks

In Chap. 2, Secs. 2.1.11 and 2.1.12 we have quoted the most relevant theoretical and computational methods for the analysis of atomistic properties of many-particle systems. For the determination of the electronic properties, other methods are relevant (methods of quantum chemistry, Hartree–Fock approximation, and density functional formalism), which have not been quoted here.

These theoretical methods used so far are not nanospecific; they have been taken over from other disciplines. However, here we have to be careful. In general, the usual theoretical (computational) methods have to be developed further at the nanolevel, and it is of particular importance to recognize the weak points of these usual methods when they are applied at the nanolevel.

But what are the modifications with respect to nanoscience? What about the theoretical (computational) preparation under nanoscience conditions? This is always dependent on the system under investigation; nanosystems demand new steps and efforts in their theoretical preparation, at least in some cases.

One of the essential points in connection with nanomaterials is that their properties can respond or switch according to external conditions, for example temperature. This means that the properties of those nanomaterials have to be described as a function of temperature. Furthermore, the effects in connection with metal particles are strongly dependent on temperature. However, in the literature the electronic properties are exclusively calculated in terms of the density functional formalism (DFT) and time-dependent DFT, but these approaches only deliver zero-temperature facts. The Mermin method[26] allows one to treat the properties of an electron system for nonzero temperatures and should therefore be used in the future for nanosystems.

Atomistic simulations need precise knowledge of interaction potentials as input, and it is in most cases not simple to find reliable interaction potentials. However, we know that nanoproperties are very sensitive to small variations in the potential used in the calculation.

In many cases the determination of potentials has been done in an uncontrolled manner. The modeling of potentials has often been done on the basis of relatively drastic simplifications. This loss of precision is often compensated by further inaccuracies in connection with other parameters of the system. In other words, within the frame of such a treatment a certain inaccuracy is compensated by another.

The future challenges with regard to complex nanosystems should not be based on this kind of inaccurate treatment and, furthermore, the theoretical (computational) methods should not simply be adopted from other disciplines without checking for their applicability at the nanolevel. Controlled steps are necessary in connection with the theoretical and computational preparation of nanosystems, with respect to not only atomistic simulations but also zero-temperature approaches (DFT and time-dependent DFT).

Reliable theoretical analysis of nanosystems inevitably means that the theoretical (computational) methods used in the description of nanosystems have to be precisely under control (with regard to approximations and simplifications). The temperature effects and the fact that nanoproperties are very sensitive to small variations in the interaction potential require that.

Nanoclusters are important building blocks. All the effects in connection with temperature and interaction potential discussed here are of particular relevance to nanoclusters. To study clusters at zero temperature is not the relevant thing about clusters. Even at very low temperatures the atomic and molecular structure of nanoclusters can be quite different from the structure at zero temperature. Since electronic properties of nanosystems are sensitive to small variations in the structure, the electronic properties should also be calculated as a function of temperature; zero-temperature data will probably deliver reliable descriptions only for a few systems. However, in the literature we exclusively find zero-temperature studies for the determination of electronic properties.

7.3.2 On the Nanotechnological Changes of Brain Functions

The manipulation of atoms, molecules, etc. will allow us to construct new technological worlds and will bring fundamentally new possibilities in the field of medicine, biology and, in particular, with respect to brain research.

Just in the case of nanobiotechnology, big changes are expected already in the near future. It has been speculated that through nanotechnology our bodies will be transformed into illness-free, undecaying systems of permanent health. Moreover, it has been prognosticated that it will take approximately 30–50 years to develop nanotechnological means for the creation of superhuman intelligence. Here nanotechnological changes are planned and have already been started for animals.

We asked the following question: Are the theoretical tools developed so far sufficient for understanding this kind of experimental manipulation? Are we really in a position to describe brain phenomena? We discussed this point critically, by means of basic principles.

Do we really know what a brain is, and how it works? In other words, is a human able to understand his own brain? Here Gödel's theorem comes into play, and it says that no mathematical system can make complete statements about itself.

When we transfer this theorem to the brain situation, we come to the result that a human can never completely understand his own brain. Such a self-related analysis is obviously not completely possible.

We know a lot about the human brain. The problem is that we do not know what we do *not* know about it. In conclusion, we have to be careful when we "play" with the human brain. Due to Gödel's theorem we cannot completely know what we do when we change the brain functions.

There are further points that we should know when we try to change brain functions. What kind of things are changed when we manipulate brain functions nanotechnologically? Here the following question is of particular relevance: Is a human being able to recognize basic, objective reality? What kind of reality is in front of and around him? We pointed out that it cannot be the basic reality, but that what we have in front of and around us is dependent on the biological structure of the observer: it is a species-dependent reality. Why is nature organized in this way? The answer is given by the principles of evolution.

The strategy of nature is to take up as little information from the outside world as possible. Reality outside is not assessed by "true" and "untrue" but by "favorable to life" and "hostile to life." This is probably the most important idea with respect to the phenomenon of evolution.

In conclusion, in nature it is not cognition that plays the most important role but the differentiation between "favorable to survival" and "hostile to survival," at least in the early phases of evolution. With respect to this fundamental point, it is important that the members of a species are able to recognize and to assess the earliest possible changes in the environment within the framework of a consistent world view, in particular within the reality in front of and around a member.

The possibility of a fallacy has to be ruled out as far as possible. There may be no doubts concerning the particular status of the environment. For this purpose, the reality in front of a biological system, which has been designed through evolution, has to be correct but it may only contain, for economic reasons, information which is absolutely necessary for survival. Everything else in unnecessary; it encumbers and is therefore hostile to life.

The "species-dependent world view" does not have to be complete and true (in the sense of precise reproduction with respect to basic, species-independent reality) but it must be reliable and restricted, at least during the early phylogenetical phase. These are the criteria that guarantee optimal chances of survival.

All that has to be kept in mind when we plan nanotechnological changes within the frame of biological systems, in particular when we try to manipulate specific brain functions. Since the world views are species-dependent and extremely relevant to survival, we in general also change the world view when we manipulate brain functions nanotechnologically. In other words, in such cases we change everything that has been developed by evolution during the entire past and not only what has been developed in the last phases.

From the point of view of evolution, we may state that the impressions in front of us do not reflect basic, objective realty, which exists independently of the observer. But what we see in front of and around us reflects merely an appropriate "species-dependent reality" that is formed by the individuals from certain information from the outside world. Nanotechnological changes with respect to the brain functions will in general also change the world view, i.e. this species-dependent reality, which is essentially relevant to survival of any biological individual.

This is of course extremely important when we try to manipulate the brain nanotechnologically, because we change in general the world view of a species and, therefore, we could change the imagination of what the biological system needs for survival. Here we have to be careful.

Even when we invent nanotechnological methods for enhancing the intelligence of a human being, these changes could in principle modify simultaneously the world view of the observer that he needs for survival. In Chap. 2, Sec. 2.3.1 we have discussed a specific nanotechnological experiment for increasing mousy intelligence, and similar experiments are explicitly planned for man. This is interesting, but what do we change simultaneously? Such and similar questions have to be investigated with the help of reliable theoretical models *before* we start such experiments.

7.4 OUTLOOK

The present basic laws of physics have been formulated within the frame of the container principle (see in particular App. B and Fig. 14). Within this principle it is believed that the impressions before and around us are identical with basic, objective reality.

However, the chick experiment and the investigations by Lashley, Pribham, and Von Foerster (Chap. 2) teach us that the projection principle should be valid, and this is because the structures outside should not be identical with the structure in the picture (identical with the species-dependent world view). In other words, when we apply the container principle in the formulation of the laws of physics, we do not have the laws for the reality outside but the laws of the species-dependent reality.

Then, the following point is relevant: when we develop and change nanosystems, for example the human brain, we need a realistic conception of reality. The brain functions work in accordance with the projection principle. However, our traditional laws of physics work as if everything were embedded in space and time (the container principle).

In conclusion, the traditional physical laws are based on the container principle. There is obviously a mismatch between the description and what really happens. Without doubt, we have to be careful when we change the world around us by nanotechnological manipulations.

7.4.1 Time Within Projection Theory

We have also outlined that the self-organizing processes are presently described by equations, which have the form of Schrödinger's equation, $i\hbar\partial\Psi_{A?}(\mathbf{r}, \tau)/\partial\tau = -\hbar^2(2m_0)\Delta\Psi_{A?}(\mathbf{r}, \tau) + V_{A?}(\mathbf{r}, \tau)\Psi_{A?}(\mathbf{r}, \tau)$ [Eq. (8)]. Moreover, we have pointed out in Chap. 6 that the time τ in this equation plays the role of a classical quantity, and this is a deficiency. Just in the quantum-theoretical description of self-organizing processes, the time is an important quantity and should also be a quantum-theoretical quantity, but this is not the case. We have discussed this point in Chap. 6.

In conclusion, with regard to Eq. (8) there are two huge drawbacks:

(1) We work within the container principle,
(2) The time behaves classically.

It is astonishing that the application of the projection principle automatically leads to a "quantum time." Details of this projection theory are given in Refs. 13–15.

What does "quantum time" mean within projection theory? Here we have a system-specific time t, which is a new variable and is not known in conventional physics. For t we have a probability distribution $\{t\}$. This distribution has the following characteristic: at time τ, one of the possible values for t can be observed with a certain probability. That is to say, at time τ one value t of the time variable can in principle lie in the interval between $-\infty$ and ∞ : τ : $-\infty \leq t \leq \infty$. In other words, at time τ the system-specific time fluctuates arbitrarily between the past and the future.

The quantity τ is the reference time. In order to be able to compare the various time structures, produced by the various systems under investigation, we need a reference time structure, which can be chosen arbitrarily but should be the same in all laboratories. For example, the reference time structure may be measured by the clocks used in everyday life. This reference time goes strictly from the past to the future and never goes backward. This arrow of time has been fixed according to the time perception of living human bodies. For this time we have used so far the Greek letter τ.

We may summarize the situation in projection theory as follows: At time τ we have probability distributions $\{p\}$, $\{r\}$, $\{E\}$, $\{t\}$ for the momentum p, the position r, the energy E, and the system-specific time t.

In the usual, traditional quantum theory there is no system-specific time t definable. Thus, instead of τ : $\{p\}$, $\{r\}$, $\{E\}$, $\{t\}$ (projection theory), we have here τ : $\{p\}$, $\{r\}$, E. There is no probability distribution for the energy E, but E takes at each time τ definite values.

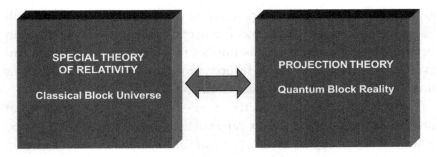

Fig. 27 The classical block universe of the special theory of relativity and the quantum block reality of projection theory.

In the case of classical mechanics, we have at time τ no probability distributions for the variables $\mathbf{p}$, $\mathbf{r}$, and $E : \tau : \mathbf{p}, \mathbf{r}, E$.

7.4.2 Quantum Block Reality

It is remarkable that within projection theory the system-specific time t fluctuates at time τ arbitrarily between the past and the future $(\tau : -\infty \leq t \leq \infty)$. We have here a "quantum block reality" with respect to space and time. This kind of block reality or block universe is new within the frame of quantum phenomena but not when we consider the special theory of relativity, which is classical in character. Here, too, we have a block universe (see also Fig. 27).

The situation within the special theory of relativity is excellently characterized in Ref. 29. John Wheeler's statement[29] "Time is nature's way of keeping everything from happening at once" is very close to that we obtained within the frame of projection theory about the nature of time. His statement is an interpretation of the situation given within the special theory of relativity. This is well summarized in Ref. 29: "Minkowski's 4D spacetime is often referred to as the block universe model. Once time is treated like a fourth dimension of space we can imagine the whole of space and time modelled as a four-dimensional block.... Here we have a view of the totality of existence in which the whole of time — past, present and future — is laid out frozen before us. Many physicists, including Einstein later in his life, pushed this model to its logical conclusion: in 4D spacetime, nothing ever moves. All events which have ever happened or ever will happen exist together in the block universe and there is no distinction between past and future. This implies that nothing unexpected can ever happen. Not only is the future preordained but it is already out there and is as unalterably fixed as the past."

This behavior is also reflected in $\tau : -\infty \leq t \leq \infty$ (the property of projection theory), because the range of the time interval $-\infty \leq t \leq \infty$ is statistically occupied in the course of time τ and we cannot agree on where the present should be, just as in the case of the block universe of the special theory of relativity.

Appendix A

MOLECULAR DYNAMICS

A.1 BASIC PRINCIPLES

The relatively strong anharmonicities in nanosystems are not negligible even at low temperatures, and we have to describe such systems on the basis of the most general formulation. In the case of classical systems consisting of N particles (atoms, molecules, and ions), this can be done on the basis of Newton's equations of motion if the interaction potential between the N particles is known. In most cases we may use pair potentials, i.e. if many-body interactions can be neglected. Such a general description can be achieved within the framework of molecular dynamics calculations.

Within molecular dynamics Newton's equations of motion are solved by iteration with the help of a high-speed computer, and we obtain the coordinates $r_i = (x_i, y_i, z_i)$ and the momenta $p_i = (p_{xi_i}, p_{yi_i}, p_{zi_i})$ $(i = 1, \ldots, N)$ of all N particles as a function of time τ. This is the total information of the system under investigation. In particular, the anharmonicities are treated without approximation. However, the solutions to Newton's equations of motion require initial values for the coordinates and momenta (velocities). The temperature is also fixed by the initial conditions for the momenta.

In other words, the general description of the properties of classical many-particle systems (for example classical nanosystems) can be obtained by means of molecular dynamics calculations without the use "simple models" and other simplifying assumptions. Such simple models and simplifying assumptions, respectively, are not known at the microscopic nanolevel and can only be introduced in a phenomenological or empirical way. In contrast to molecular dynamics, no analytical model introduced so far has been able to cover the whole complexity of such systems.

A.2   AVERAGE VALUES

Imagine a space of $6N$ dimensions whose points are determined by the $3N$ coordinates $s = (q_1, q_2, \ldots, q_{3N})$ and the $3N$ momenta $p = (p_1, p_2, \ldots, p_{3N})$. This space is the so-called phase space, and each point at time τ corresponds to a mechanical state of the system. The evolution of the system as a function of time is completely determined by Newton's equations of motion if the system behaves classically; this collective motion is described by a trajectory in phase space [see Fig. A1(a)]. The trajectory passes through the element $dsdp$ at point (s, p) of the phase space, and the points in Fig. A1(b) indicate how often the elements of the phase space have passed through by the trajectory given in Fig. A1(a).

In other words, instead of the trajectory [Fig. A1(a)] we have now a "cloud" of phase points. The cloud is a big number of systems of the same nature, but differing in the configurations and momenta which they have at a given instant. In summary, instead of considering a single dynamic system [Fig. A1(a)], we obtain various systems, all corresponding to the same set of equations of motion (Hamiltonian). This collection of systems is the so-called statistical ensemble. This is the system under investigation in one of its possible states. The introduction of the statistical ensemble is very useful regarding the relationship between dynamics and thermodynamics.

Now we may define averages with respect to the statistical ensemble [Fig. A1(b)] or to the trajectory [Fig. A1(a)]. Again, the ensemble is defined by all possible states which the system under investigation can take on; in the case of the trajectory, the system moves through phase space in the course of time τ.

Let us consider the velocity $v(\tau)$ of a single particle of a monatomic system with N particles. The probability dW of finding the velocity

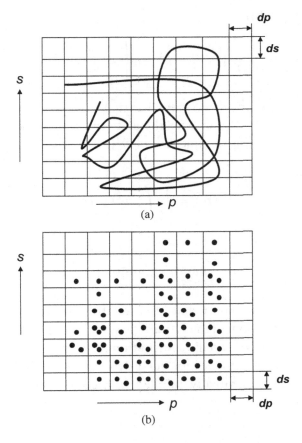

Fig. A1   (a) Trajectory in phase space, (b) statistical ensemble. For details, see the discussion in the text.

of a single particle in the interval v, $v + dv$ is expressed by Maxwell's distribution,

$$g(v) = \frac{4}{\sqrt{\pi}} v^2 \left(\frac{m}{2k_B T} \right)^{3/2} \exp \left\{ -\frac{mv^2}{2k_B T} \right\}, \qquad \text{(A1)}$$

and for the probability we get $dW = g(v)dv$. m is the mass of the particle, T the temperature of the system, and k_B Boltzmann's constant.

Then, the mean square velocity $\langle v^2 \rangle$ is given by averaging of v^2 over all members of the ensemble [over all possible states of the single particle; Fig. A1(b)]:

$$\langle v^2 \rangle = \frac{4}{\sqrt{\pi}} \left(\frac{m}{2k_B T} \right)^{3/2} \int_0^\infty v^4 \exp\left\{ -\frac{mv^2}{2k_B T} \right\} dv = 3\frac{k_B T}{m} \quad \text{(A2)}$$

We obtain exactly the same value for $\langle v^2 \rangle$ if we average v^2 over the trajectory [Fig. A1(a)], i.e. over the state v^2, which takes the single particle as a function of time: $v^2 = v^2(\tau)$. If we define a function $v^2(\tau_0)$ by

$$v^2(\tau_0) = \frac{1}{\tau_0} \int_0^{\tau_0} v^2(\tau) d\tau, \quad \text{(A3)}$$

we must have

$$\lim_{\tau_0 \to \infty} v^2(\tau_0) = \langle v^2 \rangle \quad \text{(A4)}$$

In other words, we expect that after a sufficiently large time τ_0 the velocity $v(\tau)$ of an atom of the N-particle system has passed through all the states given by Maxwell's distribution (A1). It has turned out that within the frame of realistic molecular dynamics models the condition $\tau_0 \to \infty$ in (A4) is already fulfilled in a good approximation for $\tau_0 > 10^{-11}$s.

A.3 INITIAL VALUES

Again, from the solution of Newton's equations of motion we obtain the coordinates and the momenta (velocities) of all N particles as a function of time τ. This is the total microscopic information about the system under investigation. However, the solution of the equations of motion provides that initial values for the coordinates and the velocities of the N particles are available.

A.3.1 Initial Values for the Coordinates

In the case of liquids and gases, the particles can be distributed randomly with appropriate density. In the case of crystals (with and without surface), the particles are normally positioned within an array so that the perfect lattice structure, appropriate to the system under investigation, is generated. This procedure can also be chosen for nanosystems. The system is of course not fixed to this structure but the structure develops in the course of time until a stationary state is reached.

A.3.2 Initial Values for the Velocities

When there are no external forces acting on the system, the directions of the velocities $v_i/|v_i|$, $i = 1, 2, \ldots, N$, at the initial time should be distributed randomly so that the sum over $v_i/|v_i|$, $i = 1, 2, \ldots, N$, must be zero or constant, and this must be valid for all times τ; this condition is necessary because the conservation of momentum must be fulfilled at each time τ during the molecular dynamics calculation.

In thermal equilibrium the magnitudes of the particle velocities are distributed according to Maxwell's distribution (see in particular App. A, Sec. A.2). It is, however, more convenient to choose initially for all particles the same magnitude of velocities. Due to the mutual interaction of the N particles, the distribution for the velocities develops in the course of time toward Maxwell's distribution.

In other words, the probability $dW = g(v)dv$ of finding the velocities between v and $v + dv$ is initially expressed by a delta function, and this means that the system is initially not in thermal equilibrium. With the help of the function

$$\alpha(\tau) = \frac{\frac{1}{N} \sum_{i=1}^{N} \left[v(\tau)_i^2 \right]^2}{\left[\frac{1}{N} \sum_{i=1}^{N} v(\tau)_i^2 \right]^2}, \tag{A5}$$

where $v_i(\tau)$, $i = 1, 2, \ldots, N$, are again the velocities obtained from the molecular dynamics calculation, we can study at which point in time Maxwell's distribution is reached. In the case of Maxwell's distribution [see Eq. (A1)], $\alpha(\tau)$ takes exactly the value $\alpha(\tau) = 5/3$ for all times τ. With our initial velocity distribution (all velocities have the same magnitude), we get $\alpha(\tau) = 1$.

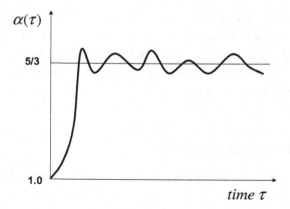

Fig. A2   Schematic representation of $\alpha(\tau)$. The system is initially not in thermal equilibrium [$\alpha(\tau) \neq 5/3$]. After a few hundred molecular dynamics time steps, thermal equilibrium (Maxwell's distribution) is reached. Because of the finite number of particles, the system fluctuates around 5/3.

It can be demonstrated by realistic molecular dynamics calculations that thermal equilibrium [$\alpha(\tau) = 5/3$] is reached after a few hundred time steps if we start from $\alpha(\tau) = 1$; the time step is of the order of 10^{-14} s. Because of the finite number of particles, $\alpha(\tau)$ fluctuates around its equilibrium value of 5/3 (see also Fig. A2); these fluctuations get small with increasing particle number N and are physically realistic. Clearly, in the case of macroscopic systems ($N \to \infty$) such fluctuations are zero.

A.3.3   Temperature of Molecular Dynamics Systems

The temperature of molecular dynamics systems is dependent on time τ and is defined by the well-known relation

$$\frac{1}{2}m\langle \mathbf{v}(\tau)^2 \rangle = \frac{3}{2}k_B T(\tau). \tag{A6}$$

With

$$\langle \mathbf{v}(\tau)^2 \rangle = \frac{1}{N}\sum_{i=1}^{N}\mathbf{v}_i^2(\tau) \tag{A7}$$

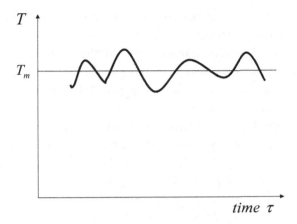

Fig. A3 The temperature as a function of time (schematic representation). T_m is the mean temperature. Because of the finite number N of particles, the temperature fluctuates around T_m.

we get for the temperature

$$T(\tau) = \frac{m}{3Nk_B} \sum_{i=1}^{N} \mathbf{v}_i^2(\tau). \tag{A8}$$

As in the case of $\alpha(\tau)$, the temperature $T(\tau)$ fluctuates around the mean temperature T_m (see Fig. A3).

These temperature fluctuations are not artificial but actually appear in the case of small units with a finite number of particles, such as nanosystems. In other words, temperature fluctuations are quite natural. The fluctuations can be used in particular for the calculation of material properties, for example the specific heat at constant volume.

A.3.4 Measurable Quantities

As has been mentioned several times, the microscopic information from molecular dynamics calculations is given by the coordinates and the momenta (velocities) of all N atoms (molecules, etc.) as a function of time. This is in fact the total information about the system under investigation.

However, this total information is not directly accessible to measurements but only certain averaged quantities can be measured, in particular those that are characteristic of the structure and dynamics of nanosystems, but also for systems without surface (solids in the bulk). In Chap. 2, Sec. 2.1, we have discussed the properties of clusters made of aluminum.

Since averaged quantities are relevant in connection with measurements, the theoretical expressions should also be this kind of averages

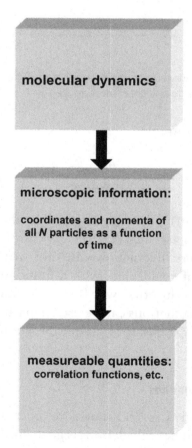

Fig. A4   The basic molecular dynamics information is given by the coordinates and the momenta (velocities) of all N particles as a function of time. From these we can directly calculate measurable quantities, such as the pair correlation function and the (generalized) phonon density of states.

and can be calculated by means of the basic molecular dynamics information — the coordinates and the momenta (velocities) of all N particles as a function of time. The pair correlation function, the triplet correlation function, and the (generalized) phonon density of states are typical examples (Fig. A.4).

A.4 SUMMARY

Within molecular dynamics Newton's equations of motion are solved by iteration with the help of a high-speed computer, and we obtain the coordinates $r_i = (x_i, y_i, z_i)$ and the momenta $p_i = (p_{xi}, p_{yi}, p_{zi})$, $i = 1, \ldots, N$, of all N particles as a function of time τ. This is the total information of the system under investigation. In particular, the anharmonicities are treated without approximation.

Within such calculations we may define averages with respect to statistical ensembles, as well as to the trajectory formed by the systems. The ensemble is defined by all possible states which the system under investigation can take on; in the case of the trajectory, the system moves through phase space in the course of time. Both possibilities can be applied within typical molecular dynamics models.

For the solution of the equations of motion, we need initial values for the coordinates and velocities of all N particles. These initial conditions can be chosen more or less arbitrarily but should be in accord with the thermodynamic state of the system under investigation.

We also need a realistic interaction potential; more details are given in App. B, particularly B.6. The interaction potential has to be constructed very carefully.

Due to the finite number of particles, molecular dynamics systems fluctuate around Maxwell's distribution for velocities and the temperature also varies in the course of time. All these fluctuations are not artificial but natural.

Appendix B

INTERACTION

B.1 BASIC PRINCIPLE: INTERACTING SPHERES

The mathematical description of physical phenomena started with Isaac Newton more than 300 years ago. Newton introduced the idea of force, i.e. a force which acts between material bodies. He was interested in stars and planets. These were the bodies to be considered. We regard stars and planets as approximately smooth spheres.

It is most astonishing that the essential characteristics of this basic concept have been transferred to all areas of physics. This basic principle can be formulated in a somewhat generalized form: We have spheres of a given diameter which influence each other by certain forces.

As mentioned, this basic model is applied to the solution of problems in practically all disciplines of physics, i.e. from the planet system down to the area of the elementary particles, whereby the range of dimensions here goes from 10^{15} cm (planet systems) to 10^{-12} cm (atomic nuclei). Here the open parameters are only the diameters of the material bodies and the forces between them. Herwig Schopper said in connection with the elementary particle principle (here we will not discuss strings, etc.)[23]: "If we ask about the eternal, everlasting, indestructible elements in physics, then the answer would until recently have been: There are final, impenetrable building blocks of matter. The forces permit the joining of these building blocks in various ways and thus cause the continual changes in nature. This conception has endured in principle for several centuries, although certainly the question of what are to be considered the final building blocks is an issue which has been exposed to considerable change, as is also the case with our knowledge about the forces. Nevertheless — despite all progress brought about by the theory of relativity and quantum theory — the description of nature is in the last analysis based upon the materialistic idea of hard, smooth, tiny spheres with a kind of spring between them."

Schopper himself regards this naïve picture, for the following reason, as obsolete: because of the forces which hold the building blocks together, a cohesive energy between the particles appears. In connection with quarks this cohesive energy becomes comparable to their rest energy, so that it really no longer makes sense to consider quarks as independent units.

But that is of course valid not only for quarks, but for all units which are arranged in accordance with the above-mentioned basic principle. The individual particles composing the solid body are not well defined, and neither are the electrons in the atoms, or the protons and neutrons in the atomic nuclei. In principle, this even applies to the planets since between them there also must be cohesive energy. The eigenenergies of these particles and planets are of course much larger than the interaction energy. One can regard them as approximately defined, but not exactly. The principle is important here!

At any rate it is believed in traditional physics that one is able to describe all phenomena using the basic principle of "interacting spheres", and that this applies to everything, from the planet system down to the quark model.

In the case of strings etc., the interaction comes into play as an unavoidable consequence through string topology. Fictitious forces do not have to be introduced. However, here we do want to discuss the details, because in nanoscience and nanotechnology strings and branes are not the relevant factor.

B.2 BASIC CHARACTERISTICS

There is an interaction between the building blocks of matter, and this interaction can be traced back to four basic forces: strong interaction, weak interaction, electromagnetic interaction, and gravity.

These four basic forces differ in their intensity and their range. Gravitational force and electromagnetic force have an infinite range; weak interaction, by contrast, has a range of only about 10^{-15} cm, which is approximately 1/100 that of strong interaction.

Because of the short range of the strong and the weak interaction, the effects to which they lead cannot be directly perceived in everyday life. These forces are not directly accessible to our senses. In everyday life, only effects due to electromagnetic interaction and gravity are evident. The composition of condensed matter (solids, biological systems, etc.), particularly nanosystems, is practically only determined by electromagnetic interaction.

In modern physics the interaction between two particles can be imaged as follows: the description of the interaction within quantum field theory shows us a process in which the two units exchange a virtual particle, "virtual" meaning that this third particle cannot be observed.

Here we will not discuss details, but let us nevertheless make some remarks regarding the basic principle, which is based on so-called Feynman diagrams.

B.2.1 Feynman Diagrams

Processes in the submicroscopic area are described within the usual framework of physics by Feynman diagrams. One assumes that the processes which are described by such diagrams really take place in this way in nature. The basis of these diagrams is the quantum field theory, which leads to a general and important conclusion[23]: "All of the interactions in nature arise from acts of annihilation and creation of particles at definite points in space and time. There are two important ideas here: First, all interactions involve the creation and annihilation of particles; second, these creations and annihilations do not take place over a region of space nor over a span of time, but are instantaneous and localized at points. By an 'interaction' is meant simply the influence of anything on anything else. Thus, all ordinary forces, pushes and pulls of one thing on another, are interactions. Also, the decay of an unstable particle is the manifestation of an interaction. The final particles are 'influenced' by the initial particle — they come into existence only because the initial particle was there."

Example

Two electrons with charge q^- interact in accordance with Coloumb's law (see Sec. B.3) at any instant of time with a repulsive force. This is no longer the case within the frame of quantum field theory, because the repulsion is no longer held to occur at each instant; the repulsion is accomplished here by an exchange of photons γ at certain times and at certain places in space (Fig. B1, in A and B). Such processes are portrayed in space–time diagrams, and these are called Feynman diagrams. The Feynman diagram for the mutual repulsion of electrons is given in Fig. B1. The space–time points, in which particles (the space–time positions at which the photons are created and annihilated are

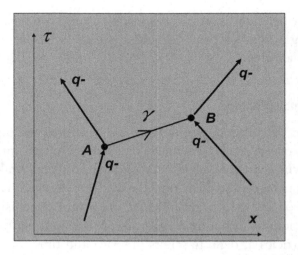

Fig. B1   Feynman diagram showing the interaction of two electrons with charge q^-. A Feynman diagram describes a process (here the repulsion of two electrons) in space and time (x, τ).

called vertices; points *A* and *B* in Fig. B1. A lot of Feynman diagrams exist, particularly with a more complex structure than that given in Fig. B1.

The aforesaid four basic forces can now be understood and described in terms of such diagrams. Let us mention the main facts:

Strong interaction

In the case of strong interactions between hadrons (such as protons and neutrons), pions are exchanged as virtual particles. For example, the interaction between a proton and a neutron is realized by a p^+ pion being exchanged, which we will not characterize further here.

The building blocks of the protons and the neutrons are the quarks. They also interact strongly and of course via a virtual particle, which is called a gluon. While pions have a rest mass, rest mass of gluons is zero.

Weak interaction

In the weak interaction between an electron and a neutrino, the so-called W boson is exchanged as a virtual particle. The rest mass of this virtual boson is different from zero. For this interaction type also, we will not go into detail here, because nanotechnology (nanoscience) is not influenced by weak interactions.

Electromagnetic interaction

This kind of interaction appears between electrically charged particles. For example (see Fig. B1), the repulsive force between two electrons comes into play via an exchange of virtual photons. It is assumed that they have a rest mass of zero.

Gravity

Gravity is by far the weakest of the four forces. Also, for gravitation the theoretical treatment leads to a mechanism which describes the interaction — as in the case of the other three basic forces — by an exchange of a virtual particle. The virtual particle itself is called a graviton; it is assumed to have a rest mass of zero.

In comparison with the three other basic forces, gravity plays a special role, which makes the whole particle concept somewhat uncertain. The reason is discussed extensively in the literature.

Gauge theories

The goal of contemporary elementary particle physics is to find for all basic forces one uniform theory in order to be able to explain, among other things, the differences between the forces. In the course of the quest for such a theory, some remarkable successes have been achieved in the last few decades. However, whether it will be possible to unify the four basic forces into one elemental force is still an open question. (We need perhaps a completely new idea; string theory is such an ansatz.)

What principles are presently considered to be important in the search for a uniform theory? Here the conception of symmetry is central: within these theories specific symmetries and their violation are the basis for all phenomena in nature. The theories involved are called non-Abelian gauge theories.

Mechanisms

Electromagnetic interaction, and in some cases also gravity, is of particular relevance when we enter the field of nanoscience and nanotechnolgy. However, what does electromagnetic interaction or gravity mean when we do not work within the framework of quantum field theory? What picture, for example, had Newton as he introduced the notion of gravity? Or how can we understand the mechanism of interaction when we study the forces between electrical charges without quantum field theory?

This point is relevant because there is a connection between the principal definition of the term "interaction" and what we have called the quantum-field-theoretical treatment of interacting systems. The

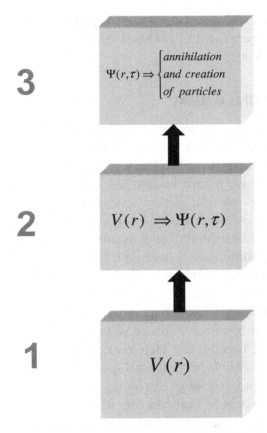

Fig. B2 Block 3: the quantum-field-theoretical description is connected with the annihilation and creation of particles (see, for example, the Feynman diagram in Fig. B1). We come to block 3 when we start from block 1 [principal definition of the notion of "interaction" $V(r)$]; from block 1 we get block 2 when we treat the physical system quantum-mechanically, leading to the wave function $\Psi(r, \tau)$. If we quantize the $\Psi(r, \tau)$ field, we obtain the quantum-field-theoretical picture.

situation is summarized in Fig. B2. Here we have three blocks: 1, 2, and 3.

We start with the principal definition of the notion of "interaction" (block 1 in Fig. B2), which is symbolized by $V(r)$, where r is the distance of the interacting systems. We get block 2 when we treat the physical system, which is based on $V(r)$, quantum-mechanically, leading to the wave function $\Psi(r, \tau)$. τ is again the time. If we quantize the $\Psi(r, \tau)$ field, we obtain the quantum-field-theoretical description (block 3). What are the roots of block 3? It is the function $V(r)$. What can we say to $V(r)$?

B.3   SOME PRINCIPAL REMARKS

The material world, as we imagine it directly, has its roots in the observations performed by the five senses in everyday life, i.e. at the level of direct and unprejudiced (assumptionless) observation.

At this level we regard bodies and matter as being *de facto* the same; bodies (i.e. matter), as they appear directly, present themselves as a continuous medium, whose boundaries in space determine their form. "Continuous medium" here means densely packed matter, as it appears in all unprejudiced (assumptionless) observations in everyday life. Within assumptionless observations reality appears to be embedded in space and time, and this impression was the basis for the container principle (see Chap. 3 and Fig. 14).

However, as has been outlined in Chap. 2, this idea has proven to be deceptive, because the facts that we have directly in front us are not independent of the biological species; it is a species-dependent reality and not the basic, objective reality. In connection with the chick experiment, it has been outlined that man and the turkey obviously have visual impressions of a chick that are fundamentally different from each other. However, so far traditional physics has not incorporated these important facts from behavior research. Therefore, when we talk in the following about the notion of "interaction," we talk about the traditional view. Within the traditional view the material world is embedded in space and time (see in particular Fig. 14).

B.3.1   The Electric Field

What gives the bodies their firmness and form is the electric forces and fields, respectively, within the atoms and between the atoms. In the atomic concept it is precisely these fields which constitute the appearance of a body familiar to us, but precise measurements also support this view.

As has been ascertained in Ref. 12 the electric field, which holds the atoms and matter together, has an infinite set of data and is therefore not accessible to actual experience, i.e. the electric field and the field idea in general have to be regarded as metaphysical. This statement has to be considered as negative when we assume that the material world is embedded in space and time. However, we will not deepen this point here.

In order to explain the nature of the electric field more fully, it is necessary to first make a few principal remarks about the quantitative description of this phenomenon.

B.3.2 Description Elements

Let us consider an atomic nucleus with mass M and an electric charge Q, which is surrounded by an electron with mass m and charge $q = -Q$. What can we say about the interaction, i.e. the mutual influence of the two charges?

The interaction between these point charges proceeds according to Coulomb's law, which formulates the force F as a function of the two charges Q and q and the distance r between them; this relation is given by Qq/r^2. F is influenced by the charge and likewise by the point r (position of the charge q). An entity which characterizes the position alone is the electric field $E(r)$; with $F = qE$ we obtain the expression Q/r^2 for it.

If the distance r between the charges q and Q is varied, "work" must be carried out. That is to say, the energy between the two charges is altered. In other words, for the two charges q and Q, separated by the distance r, there exists a potential energy $V(r)$, which is expressed by qQ/r. $V(r)$ is simply called "potential."

The charge Q is thus associated with a field $E(r)$, which acts as force on the charge q at the position $r = r(x, y, z)$. The field $E(r)$ [likewise the potential energy $V(r)$] has here a set of values like the mathematical continuum. The coordinates x, y, z have — in accordance with the mathematical continuum — an infinite set of data, and thus the entities E and V are also characterized by an infinite set of values, i.e. they are not accessible to experience; only a finite set of data is observable for a human observer. This statement is generally valid for all fields in the physical world, and applies also to the gravitational field.

But it is not only the infinite set of data that the concept of a field is a metaphysical element. A further factor is that the basic characteristics of a field cannot be experienced. We will show in the next section, when we talk about "action at a distance" and "proximity effect," that it becomes entirely a matter of belief whether an electric field exists in reality or not. The reason is simple: there exists an equivalent idea, known as "action at a distance," which is however, completely different from the field conception (field effect).

B.3.3 "Action at a Distance" and the "Proximity Effect"

What is an electric field? A field is a kind of space-filling "medium" whose infinite set of data exists simultaneously.

Can we prove this space-filling medium experimentally? Does it produce, in other words, a reading on a measuring instrument which can be directly attributed to the medium "field"? The answer is negative because in the measurement of E one proceeds as follows, whereby it will be assumed at first that such a field actually exists.

An electric charge, which we call q_m, produces around itself an electric field E_m, which for symmetry reasons has no force effect on the producing charge q_m; the resulting force is zero. Thus, for the proof of E_m a second charge q_t is needed, which experiences a force in the field E_m. Now the second charge q_t, which is necessary for the proof of the field E_m of the first charge q_m, also produces around itself a field E_t, which is superimposed onto the other field E_m. The original symmetry of the field to be measured is therefore disturbed by the field of the second charge, and a force effect is produced. This force does not depend therefore on the properties E_m alone but also on the field E_t, the eigenfield of q_t. One can avoid this difficulty by making the electric charge and the dimensions of the second body as small as possible, in order to decrease the influence of E_t below each limit. If q_t fulfill this condition, it is called "test charge." The force must then be measured with this test charge as a function of r, in order to be able to determine the electric field E_m of the charge q_m.

One measures therefore a force at position r at a certain time τ and not the diversity of the field, because one cannot conclude from the effect of the force at position r that there is actually something simultaneously at the other places in space, i.e. the field as a space-filling medium. One can of course make many measurements, but not infinitely many.

In accordance with this measuring process, the field idea therefore becomes entirely a matter of belief. But even when one can measure an infinite number of values, the existence of a field remains a matter of belief, since the force effect on the test charge can also be explained without using the field effect, i.e. by what is known as "action at a distance."

The idea of action at a distance came into play in connection with Newton's gravitational law. Let us first discuss Newton's basic conception, after which we will talk about the forces.

B.3.4 Newton's Equations of Motion

The aim of Newton's equations is to establish laws governing the trajectories of moving bodies. Apart from space, characterized by the coordinates x, y, z and time τ, a further element of description is, according to Newton, the (inert) mass m. If the body is considered as a mass point, it is characterized at time τ by the mass m and the coordinates $\{x(\tau), y(\tau), z(\tau)\}$. If we have N bodies, we have at time τ the coordinates $\{x_i(\tau), y_i(\tau), z_i(\tau)\}$ and the masses $m_i, i = 1, 2, \ldots, N$.

According to Newton, for these N masses there are N equations of motion. It is a system of coupled differential equations. This makes it possible to determine the trajectories (paths) for all N masses. In order to be able to solve the equations, the coordinates and velocities of all N masses must be known at any given time τ_0. In other words, we need initial values for the coordinates and velocities.

Newton's equations of motion include the forces which act between the bodies. In the case of $N = 2$ we have two masses m_1 and m_2, having the distance r, and the force is proportional to

$m_1 m_2 / r^2$ (gravitational law). If the distance between the two masses is altered, "work" must be done, i.e. the energy between the masses varies. In other words, for the two masses m_1 and m_2, with the distance r between them, there exists a potential energy $V(r)$, which is directly proportional to $m_1 m_2 / r$.

The equations of motion as well as the gravitational law are the basics of Newton's mechanics, which was unusually successful and is still used extensively. In particular, the concrete mathematical formulation of the gravitational law made it possible to compare theoretical statements with experimental data.

B.3.5 Action at a Distance

The forces have the effect that the N bodies influence each other and are functions of the distances between them. The force, which acts at time τ on body i, is therefore determined by the positions of all the $N-1$ other bodies which interact with i. According to Newton each of the $N-1$ bodies therefore momentarily exerts a force on body i; any change in the position of the $N-1$ bodies is felt by body i without any delay (retardation), i.e. completely independent of the distance which the other bodies have from body i.

According to Newton's mechanics, it appears to be the case that the forces between the masses m_1 and m_2, m_1 and m_3, ... act across space instantly. This is because in the gravitational law there is a relationship between the spatially separated positions of the masses and no intermediate position appears in the law. This suggests that gravitational forces work at a distance, i.e. the interaction comes about through an action-at-a-distance effect: the gravitational forces reside in body i, but come into effect at the location of the other $N-1$ bodies. The space between the bodies is free of gravitation in this view.

Newton himself opposed this idea of forces acting at a distance. This restraint was soon given up by his successors, partly because of the success of Newtonian mechanics, but also because of the unsuccessful search for a mechanism. This point will be discussed in more detail below.

B.3.6 Comparison of Gravitational Force and Electrostatic Force

Coulomb's law Qq/r^2 of electrostatic force, which we have introduced above, has the same mathematical form as the gravitational force $m_1 m_2/r^2$. At the time Coulomb formulated the law Qq/r^2, this formal agreement had the consequence that the forces acting between electric charges were also interpreted as forces acting at a distance. On the other hand, Newton's gravitational law can be constructed and formulated as a field effect law (the "proximity effect").

In fact, the proximity effect (based on a space-filling medium — the field) and action at a distance are equivalent conceptions which do not differ either mathematically or experimentally.

B.3.7 Summary

The field effect is a proximity effect, since the force effect on the charge q (Fig. B3) is exactly determined by the electric field at the position where the charge is also localized; other positions do not play any part. The field itself can be understood as a kind of space-filling medium.

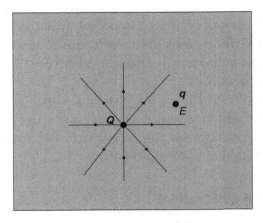

Fig. B3 In the case of the "proximity effect," it is believed that the space is filled with a "medium" — the field. The force effect on the charge q is determined by the electric field E at the position where q is located. Hence the designation "proximity effect."

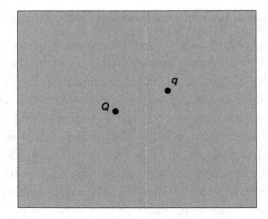

Fig. B4   In the case of "action at a distance," the space is filled with charges (q and Q in the case discussed here); otherwise the space is empty. The force felt by q and coming from Q acts across the space instantly.

In the case of action at a distance, the space contains only point charges and nothing else. The force which is felt by the charge q emanates from the charge Q and leaps across space in an instant (see also Fig. B4). As mentioned, the space between and around the charges is empty.

The proximity effect and action at a distance, so qualitatively different as they may be, are nevertheless exactly equivalent. The two conceptions are described by the same mathematical formulas and, furthermore, it is not possible to make a choice between them, since there is no experimental way to distinguish between them as they have the same consequences.

In other words, for the specific elements, which are typical of the proximity effect and action at a distance respectively, there are no element-specific readings on any measuring instrument. Thus, both ideas must be regarded as metaphysical.

B.4   IN SEARCH OF A MECHANISM

Newton's law (or Coulomb's law) tells us what happens, but not why it happens. Where does the gravity or the electric field come from? What are the mechanisms behind these forces?

The proximity effect and action at a distance are merely expressions or interpretations of the gravitational law (or, similarly, the force law which exists between charges). The force law $m_1 m_2 / r^2$ cannot, however, be derived from these notions.

Many people believe that a mechanism which is composed of many familiar single processes (preferably from everyday life) can explain the mathematical structure of the gravitational law $m_1 m_2 / r^2$. What mechanism is, for example, responsible for the fact that the forces expressed by $m_1 m_2 / r^2$ are inversely proportional to the square of the distance between the masses? As already mentioned, the proximity effect and action at a distance cannot give an answer to this question, since they interpret the relation $m_1 m_2 / r^2$ but are not the source of its derivation.

B.4.1 Application of the Particle Picture

A beautiful example of a mechanism used in an attempt to explain the gravitational law has been given by Richard Feynman. Since this point is of particular importance and, furthermore, since Feynman's formulation is brilliant, let us reproduce the original text[24]:

"Suppose that in the world everywhere there are a lot of particles, flying through us at very high speed. They come equally in all directions — just shooting by — and once in a while they hit us in a bombardment. We, and the sun, are practically transparent for them, practically but not completely, and some of them hit. Look, then, what would happen" (Fig. B5).

"S is the sun and E the earth. If the sun were not there, particles would be bombarding the earth from all sides, giving little impulses by the rattle, bang, bang of the few that hit. This will not shake the earth in any particular direction, because there are many coming from one side as from the other, from top as from bottom. However, when the sun is there the particles which are coming from that direction are partly absorbed by the sun, because some of them hit the sun and do not go through. Therefore the number coming from the sun's direction towards the earth is less than the number coming from the other sides, because they meet an obstacle, the sun. It is easy to see that the

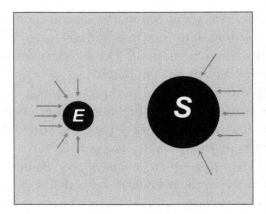

Fig. B5   Can the attraction between the earth E and the sun S be explained on the basis of impinging particles? The answer is no!

farther the sun is away, of all the possible directions in which particles can come, a smaller proportion of the particles are being taken out. The sun will appear smaller — in fact inversely as the square of the distance. Therefore there will be an impulse on the earth towards the sun that varies inversely as the square of the distance. And this will be a result of a large number of very simple operations, just hits, one after the other, from all directions. Therefore the strangeness of the mathematical relation will be very much reduced, because the fundamental operation is much simpler than calculating the inverse of the square of the distance. This design, with the particles bouncing, does the calculation."

"The only trouble with this scheme is that it does not work, for other reasons. Every theory that you make up has to be analysed against all possible consequences, to see if it predicts anything else. And this does predict something else. If the earth is moving, more particles will hit it from in front than from behind. (If you are running in the rain, more rain hits you in the front of the face than in the back of the head, because you are running into the rain.) So, if the earth is moving it is running into the particles coming from toward it and away from the ones that are chasing it from behind. So more particles will hit it from the front than from the back, and there will be a force opposing any motion. This force would slow the earth down in its orbit, and it certainly would not have lasted the three or four billion years (at least) that it has been going around the sun. So that is the end of the story. 'Well,' you say, 'it was a good one, and I got rid of the mathematics for a while. Maybe I could invent a better one.' Maybe you can, because nobody knows the ultimate. But up to today, from the time of Newton, no one has invented another theoretical description of the mathematical machinery behind this law which does not either say the same thing over again, or make the mathematics harder, or predict some wrong phenomena. So there is no model of the theory of gravitation today, other than the mathematical formula."

The text by Feynman obviously shows that gravitation cannot be explained on the basis of the "mechanical–corpuscular world view" that was the standard view of the 17th century. Newton therefore opposed his own theory because gravitation could not be described by the mechanical–corpuscular world view. Regarding this point, the following brief comment by Thomas Kuhn is relevant:

"Yet, though much of Newton's work was directed to problems and embodied standards derived from the mechanical–corpuscular world view, the effect of the paradigm that resulted from his work was a further and partially destructive change in the problems and standards legitimate for science. Gravity, interpreted as an innate attraction between every pair of particles of matter, was an occult quantity in the same sense as the scholastics' 'tendencies to fall' had been. Therefore, while the standards of corpuscularism remained in effect, the research for a mechanical explanation of gravity was one of the most challenging problems for those who accepted the *Principia* as paradigm. Newton devoted much attention to it and so did many of his eighteenth-century successors. The only apparent option was to reject Newton's theory for its failure to explain gravity, and that alternative, too, was widely adopted. Yet neither of these views ultimately triumphed. Unable either to practice science without the *Principia* or to make that work conform to the corpuscular standards of the seventeenth century, scientists gradually accepted the view that gravity was indeed innate. By the mid–eighteenth century that interpretation had been universally accepted, and the result was a genuine reversion (which is not the same as retrogression) to a scholastic standard. Innate attraction and repulsion joined size, shape, position, and motion as physically irreducible primary properties of matter."

B.5 DISCUSSION OF THE BACKGROUND

In Sec. B.3, we have discussed the situation regarding the gravitational law $m_1 m_2 / r^2$. This law could not and cannot be deduced on the basis of reasonable models. The mechanical–corpuscular world view did not lead to a useful result.

Again, Newton's laws tell us *what* happens, but not *why* it happens. Where does the gravity or the electric field come from? What are the mechanisms behind these force laws? This question could not be answered.

The proximity effect and action at a distance are merely expressions or interpretations of the gravitational law (or, similarly, the force law which exists between charges). The force law $m_1 m_2 / r^2$ cannot, however, be derived from these notions. Thus, we have no explanation for the gravitational law (Coulomb's law) but merely mathematical expressions for gravity and the forces between electrically charged masses. This situation is unsatisfactory. The main points can be summarized as follows:

(1) As the text by Feynman demonstrated, gravitation could not be explained by impinging corpuscles. He wrote, "...there is no model of the theory of gravitation today, other than the mathematical formula."

(2) Newton did not really accept the situation; his goal was to apply the corpuscular standards of the 17th century.

(3) Scientists gradually accepted the view that gravity was indeed innate. Innate attraction and repulsion joined size, shape, position, and motion as physically irreducible primary properties of matter.

However, when we take the view of modern theoretical physics we know why the corpuscular world view is not able to explain gravity. The solution of the questions formulated above (no mechanism for gravity or for Coulomb's law) is probably more basic, more sophisticated. Here we obviously touch the term "world" itself.

The "objective world" is simply not identical with what we have in front of us within our observations in everyday life, but we have to distinguish between the world we directly observe and the world which exists independently of the observer.

The chick experiment and in particular the investigations by Lashley, Pribham, and Von Foerster teach us that the structures outside should not be identical with the structure in the picture. All these points have been analyzed in Chap. 2, Secs. 2.3.4 and 2.3.5. In other words, we have a "species-dependent reality" (a picture of reality) and a "basic, objective reality" which cannot be observed by human beings and probably also not by other biological systems.

The corpuscular world view with its impinging particles does not distinguish between species-dependent reality and basic, objective reality. Just the fact that we cannot find a mechanism for the gravitational law $m_1 m_2 / r^2$ or for Coulomb's law Qq/r^2 is a proof that the imagination of what we call "reality" is too simple within the corpuscular world view. In other words, the situation is obviously more basic, more complex than thought.

Feynman's and Kuhn's discussion revealed quite clearly that a mechanism for the explanation of $m_1 m_2 / r^2$ and Qq/r^2 is not possible in connection with the space in front of us, i.e. within species-dependent reality, and this is because this space contains only geometrical positions and no real masses. Geometrical positions impinging on other geometrical positions cannot lead to physically real processes, but the formulas $m_1 m_2 / r^2$ and Qq/r^2 reflect in their original form physically real processes.

One may therefore state that the chick experiment and Feynman's analysis within the corpuscular world view lead to the same result. Since an explanation for $m_1 m_2 / r^2$ and Qq/r^2 could not be found, the corpuscular world view failed.

Let us briefly summarize the main characteristics of the corpuscular world view. It works within the following imagination: the masses are embedded in space and time, and there is nothing else; this situation reflects the basic, objective reality within the corpuscular world view.

The mass is characterized by one parameter, normally symbolized by the letter m. The motion through space leads to a trajectory, and we have at each time τ a certain velocity v leading to a momentum $\mathbf{p}$ and an energy E. However, as we have recognized above, this view cannot explain the forces between the masses. The corpuscular world view is obviously too simple; the reality, on which this world view is based, is obviously not what we have called above "objective reality."

B.5.1 Two Remarks

(1) The quantum-field theoretical phenomena (level 3 in Fig. B2) and the field effect (level 1 in Fig. B2) are based on the view that the phenomena in space and time reflect basic reality.

The law Qq/r^2 manifests itself on level 3 (quantum-field-theoretical view) by processes that are noninstantaneous and are accompanied by an exchange of virtual particles. On the other hand, the processes due to Qq/r^2 on level 1 can be considered as taking place in an instantaneous way without the exchange of corpuscles. Neither the quantum-field-theoretical effects on level 3 nor the effects (proximity effects) on level 1 are able to derive the law Qq/r^2; they merely reflect interpretations.

(2) Newton could not accept the view that gravity reflects an innate property of the masses. On the other hand, he was caught in the corpuscular world view. If he had a more realistic imagination of what is called "reality," one can be sure that physics would have taken another direction. But we are still working on this old reality conception and not on what is reflected by Fig. 15: we have species-dependent reality, and basic, objective reality is not accessible to a human being.

Just in connection with nanotechnological changes of brain function, we need a reliable situation. We need conceptions of the world that are tailor-made for the functions inside the brain. Our conception of reality must be consistent with that on which the brain functions are based; a mismatch can lead to serious problems.

B.6 FORCE TYPES WITHIN ELECTROMAGNETIC INTERACTION

Physics presently works within the container principle[15]; here everything is embedded in space and time, i.e. the space–time is considered as a container. Within this view, gravity and so on are an innate property of matter, i.e. this conception has to be considered above the corpuscular world view.

More realistic is the projection principle; here reality is projected onto space and time. The projection principle directly follows from the situation given in Fig. 15. Projection theory is just at the beginning, but a lot of details have already been worked out.[13,14] Nevertheless, in practical calculations we are still dependent on the physical laws, which have been derived on the basis of the container principle. However, we have to ask whether it is meaningful at all to describe nanosystems, particularly brain functions, in this way.

We may say that traditional physics works as if everything were embedded in space and time. In fact, in the description of the forces between material objects we use the traditional laws which are based on the container principle. Again, the brain functions work in accordance with the projection principle (the chick experiment and other findings dictate this view), indicating that we have to be careful when we try to manipulate and to describe them on the basis of the container principle.

Among all the forces of nature (strong interaction, weak interaction, gravity, and electromagnetic force), only electromagnetic interaction is relevant to systems in nanoscience and nanotechnology (Sec. B.2). Within electromagnetic interaction we have to distinguish between five force types: ionic binding, van der Waals interactions, covalent bonds, metallic binding, and hydrogen bridge bonds.

All these force types are based, in the quantitative description, on the Coulomb potential qQ/r. As said, the expression qQ/r describes the interaction between point charges. However, in general we have more or less complicated charge distributions (atoms, molecules, ions), and we have to know the interaction potentials between these charge distributions. Thus, we distinguish between ionic binding, van der Waals interactions, covalent bonds, metallic binding, and hydrogen bridge bonds. The starting point is in all cases the Coulomb potential qQ/r, and the complex interactions between complex charge distributions are derived on the basis of qQ/r. Let us briefly describe the features of these interaction (force) types. We will exclusively do that qualitatively.

B.6.1   Ionic Binding

The ionic interaction is characterized by different atoms (such as Na and Cl) which have exchanged electrons, so that the many-particle systems consist of positively and negatively charged ions (Fig. B6). The interaction law as a function of distance r between the ions is simple and is described by Coulomb's law if the distances are larger than $r_+ + r_-$, where r_+ and r_- are the radii of the positively and negatively charged particles. For distances $r < r_+ + r_-$ the interaction is repulsive, due to the overlap of the electron cores.

B.6.2   Van der Waals Interaction

The motion of the electrons around the atomic nucleus means that there is a certain probability that the atoms have at instant τ an electrical dipole moment (see also Fig. B7) — which, however, becomes

Fig. B6   Within ionic interactions, atoms exchange electrons and we have positively and negatively charged ions.

Fig. B7 The basis of van der Waals interactions is the fluctuating electrical dipole moments of the atoms. These momentary dipole moments interact with each other and we obtain an attractive potential.

Fig. B8 Covalent bonds. Here the interaction comes into existence by the fact that part of the electrons belong to several particles at the same instant.

zero when it is averaged over the time. In a many-particle system these momentary dipole moments interact with each other, leading to an attractive potential between the atoms of the system. Typical van der Waals systems are noble gases. Clearly, repulsive core effects have to be considered also in the description of the interaction potential at small distances.

B.6.3 Covalent Bonds

In the case of covalent binding, the interaction between particles is given by the fact that part of their electrons belong to several particles at the same instant (see Fig. B8). The probability of finding an electron, which is responsible for binding, is relatively large in the environment between the particles leading to the interaction between them. In most cases this process is based on the formation of spin-saturated electron pairs, where each atom contributes an electron so that the electron shells take on a noble gas configuration. Examples of substances with

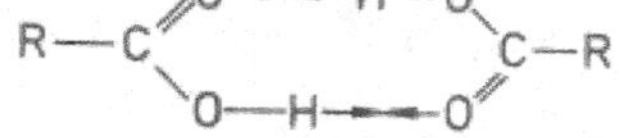

Fig. B9   Hydrogen bridge bonds. Here several molecules (atoms) are connected by hydrogen ions. In the figure the carboxylic acids form with two hydrogen bridges between the molecule groups a certain kind of double molecules.

covalent bonds are carbon, amorphous semiconductors, and hydrogen molecules.

B.6.4   Hydrogen Bridge Bonds

Several molecules (atoms) can be connected through hydrogen bridge bonds (Fig. B9). The hydrogen atom appears in such cases as mediator. For example, carboxylic acids easily form with two hydrogen bridges between the molecule groups a certain kind of double molecules (Fig. B9). Furthermore, the organic base in DNA is bonded by such hydrogen bridges; as is well known, DNA is the molecular memory of the heiress information of all life forms. In the case of the hydrogen bridge bond, two atoms do not share — as in connection with covalent binding — a electron, but a hydrogen ion (proton).

B.6.5   Metallic Binding

In the case of certain atoms (for example Al), the electrons of the outer shell are only weakly bonded. If a many-particle system is formed with such atoms, the weakly bonded electrons leave the according atoms and move through the whole system (Fig. B10). These are the conduction electrons. In other words, in metals we have the following situation: there are positively charged ions which move through the sea of conduction electrons.

In metals the ion–ion interaction is not simply given by the Coulomb potential, since the ions are screened by the conduction electrons and that makes its determination complex and complicated, respectively. Such a modeling of the ion–ion potential can be

electron gas

Fig. B10   Metallic binding. Here we have ions that are embedded within the sea of conduction electrons.

performed, for example, within the framework of pseudopotential theory.

As we have outlined, metals consist of ions of positive charge and electrons. In the case of the ionic interaction, the potential between the particles is simple and is given by the Coulomb potential. In metals the situation is more complicated because the ions are surrounded by conduction electrons (see Fig. B10); the ions are screened by the electrons, and in particular there are quantum-mechanical correlation and exchange effects among the electrons. All these facts have to be considered in the construction of ion–ion potentials $v(r)$ in metals, where r is the distance between the two ions of the system.

Within the frame of pseudopotential theory, all these effects (screening, correlation, and exchange effects) can be adequately considered. The pseudopotential method is often used for the determination of the interaction between the ions in metals. Within the frame of this theory, the problem can be treated on various levels, and a lot of pseudopotential models exist in the literature. We will not give details here.

A typical feature of the ion–ion potential $v(r)$ in metals is that the long-range part oscillates, and this because of the conduction electrons. This is schematically shown in Fig. B11 in comparison with a Lennard–Jones-type potential which describes the interaction between the atoms in noble gases; these are systems without conduction electrons. In the case of noble gases, the interaction comes into play through van der Waals forces (see Fig. B7); the induced dipole–dipole interactions lead to an r^{-6} term for the long-range part of the potential.

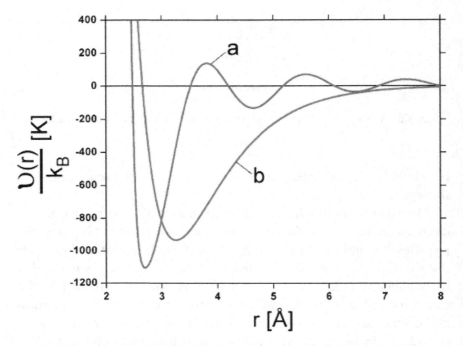

Fig. B11   Schematic representation of different potential types. (a) The screened ion–ion interaction $\upsilon(r)$ for a metal. k_B is Boltzmann's constant. Due to the screening of the conduction electrons, the long-range part of the potential oscillates. (b) The typical atom–atom potential in a noble gas; the induced dipole–dipole interactions lead to an r^{-6} term in the potential's long-range part. (©2006, American Scientific Publishers.)

B.7 SUMMARY

The mathematical description of physical phenomena started with Isaac Newton more than 300 years ago. He introduced the idea of force (interaction), i.e. a force which acts between material bodies. This basic idea has been valid up to the present day. Herwig Schopper wrote: "…despite all progress brought about by the theory of relativity and quantum theory — the description of nature is in the last analysis based upon the materialistic idea of hard, smooth, tiny spheres with a kind of spring between them."

There is an interaction between the building blocks of matter, and such interactions can be traced back to four basic forces: strong interaction, weak interaction, electromagnetic interaction, and gravity. These four basic forces differ in their intensity and their range.

Because of the short range of strong and weak interaction, the effects to which they lead cannot be directly perceived in everyday life. These forces are not directly accessible to our senses. In everyday life, only effects due to electromagnetic interaction and gravity are evident. The composition of condensed matter (solids, biological systems, etc.), particularly nanosystems, is practically only determined by electromagnetic interaction.

In modern physics the interaction between two particles can be imaged as follows: the description of the interaction within quantum field theory shows us a process in which the two units exchange a virtual particle, "virtual" meaning that this third particle cannot be observed.

In this section we started with the principal definition of the notion of "interaction" (block 1 in Fig. B2), which we have symbolized with $V(r)$, where r is the distance of the interacting systems. When we treat the physical system, which is based on $V(r)$, quantum-mechanically, we obtain the wave function $\Psi(r, \tau)$ where τ is the time. If we quantize the $\Psi(r, \tau)$ field, we obtain the quantum-field-theoretical description (see also Fig. B2). What are the roots of the quantum-field-theoretical description? It is the function $V(r)$. What can we say to $V(r)$?

We discussed the function $V(r)$ in connection with electromagnetic interactions and gravity. In the two cases we have similar expressions for $V(r)$: in the electromagnetic case we have qQ/r and in the case of gravity, it is proportional to $m_1 m_2/r$. Both interactions types have been interpreted by two conceptions: the proximity effect and action at a distance.

The proximity effect and action at a distance, so qualitatively different as they may be, are nevertheless exactly equivalent. The two conceptions are described by the same mathematical formulas and, furthermore, it is not possible to make a choice between them, since there is no experimental way to distinguish between them as they have the same consequences.

In other words, for the specific elements which are typical of the proximity effect and action at a distance respectively, there are no element-specific readings on any measuring instrument. Thus, both ideas must be regarded as metaphysical.

Newton's law (or Coulomb's law) tells us *what* happens, but not *why* it happens. Where does the gravity or the electric field come from? What are the mechanisms behind these forces? These are important questions when we intend to develop these laws further.

The proximity effect and action at a distance are merely expressions or interpretations of the gravitational law (or, similarly, the force law, which exists between charges). Therefore, the force law $m_1 m_2 / r^2$ cannot be derived from these notions. How can we derive the law $m_1 m_2 / r^2$? This law could not and cannot be deduced on the basis of reasonable models. The mechanical–corpuscular world view did not lead to a useful result.

Thus, we have no explanation for the gravitational law (Coulomb's law) but merely mathematical expressions for gravity and the forces between electrically charged masses. This situation is unsatisfactory.

However, when we take the view of modern theoretical physics we know why the corpuscular world view is not able to explain gravity. The solution of the questions formulated above (no mechanism for gravity or for Coulomb' law) is probably more basic, more sophisticated. Here we obviously touch the term "world" itself.

The "objective world" is simply not identical with what we have in front of us within our observations in everyday life, but we have to distinguish between the world we directly observe and the world which exists independently of the observer.

The chick experiment and in particular the investigations by Lashley, Pribham, and Von Foerster teach us that the structures outside should not be identical with the structure in the picture. In other words, we have a "species-dependent reality" (a picture of reality) and a "basic, objective reality" which cannot be observed by human beings and probably also not by other biological systems.

The corpuscular world view with its impinging particles does not distinguish between species-dependent reality and basic, objective reality. Just the fact that we could not find a mechanism for the gravitational law $m_1 m_2 / r^2$ or for Coulomb's law Qq/r^2 is a proof that the imagination of what we call "reality" is too simple within the corpuscular world view. In other words, the situation is obviously more basic, more complex than thought.

Feynman's and Kuhn's discussion revealed quite clearly that a mechanism for the explanation of $m_1 m_2 / r^2$ and Qq/r^2 is not possible in connection with the space in front of us, i.e. within species-dependent reality, and this because this space contains only geometrical positions and no real masses. Geometrical positions impinging on other geometrical positions cannot lead to physically real processes, but the formulas $m_1 m_2 / r^2$ and Qq/r^2 reflect physically real processes.

When we develop and change nanosystems, for example the human brain, we need a realistic conception of reality. The brain functions work in accordance with the projection principle. However, our traditional laws of physics work as if everything were embedded in space and time. In other words, the traditional physical laws are based on the container principle. There is obviously a mismatch between the description and what really happens. Without doubt, we have to be careful when we change the world around us nanotechnologically.

Appendix C

ATOMISTIC STANDARD MODELS

Theoretical materials research at the nanolevel not only needs solid state physics but, due to the strong anharmonicities and the strongly disturbed structure of many nanosystems (even far below the melting temperature of the corresponding bulk state), the basics of statistical mechanics and the theory of liquids are very often more appropriate as the basis for the theoretical description of such systems.

In such cases the standard model of solid state physics (ordered structure and harmonic approximation, here summarized by the symbol $\omega_p(\mathbf{q})$ for phonon frequencies) is not suitable, and we come to the description in terms of correlation functions and interaction potentials, which we will characterize by $g(r)$ and $v(r)$ — $g(r)$ is the well-known pair correlation function and $v(r)$ is the pair potential. Within this statistical-mechanical description, time correlations are of course included, such as the velocity autocorrelation function, whose Fourier transform has to be considered as the "generalized phonon density of states."

However, due to the strong anharmonicities, the dynamics of such systems is characterized by a broad range of different dynamical states, including small local vibrations and complex diffusion processes. This situation indicates that the development of a suitable *standard model* for the theoretical description of nanosystems will hardly be possible; even in the case of liquids in the bulk, a standard model could not be found up to now.

Therefore, the most important tool for the investigation of nanosystems is the molecular dynamics method (see Chap. 2 and, in particular, App. A) since anharmonicities are treated in this method without approximation, and this is important because the typical anharmonicities in connection with nanosystems cannot be considered as small perturbations to the harmonic approximation. No other microscopic method in condensed matter physics allows one to treat anharmonicities without approximation, but only in connection with phenomenological models.

Thus, the molecular dynamics method should be considered as the *standard method* for the theoretical description of nanosystems. On the basis of these considerations we come to the following rough classification scheme:

$$\omega_p(\mathbf{q})$$
(standard model for the solid)

$$\downarrow$$

$$g(r), \upsilon(r)$$
(statistical mechanics)

$$\downarrow$$

molecular dynamics
(standard method for nanosystems)

In nanoscience we often work on the basis of a few hundred atoms (molecules, ions, etc.). In general we may say that with decreasing particle number disorder effects and anharmonicities increase and this has an influence on the theoretical treatment. In this connection we have to state once more that the anharmonicties are not small perturbations to the harmonic approximation, i.e. in the treatment of such systems we are dependent on numerical methods since analytical models are not known for such cases.

References

■ ■ ■

1. D. Broderick, *The Spike: How Our Lives are Being Transformed by Rapidly Advancing Technologies*, (Tom Doherty Associates, New York, 2001).
2. K. Gödel, *Monatsh. Math. Phys. 38*, 173 (1931).
3. J. Milburn, *J. Comput. Theor. Nanosci.* **2**, 161 (2005).
4. W. Schommers, *App. Phys.* **A68**, 187 (1999).
5. P. von Blanckenhagen, W. Schommers and V. Voegele, *J. Vac. Sci Technol.* **A5**, 649 (1987).
6. L. Dagens, M. Rasolt and R. Taylor, *Phys. Rev.* **B11**, 134 (1975).
7. W. Schommers, C. Mayer, H. Göbel and P. von Blanckenhagen, *J. Vac. Sci. Technol.* **A13**, 1413 (1995).
8. W. Schommers, *Z. Phys.* **B24**, 171 (1976).
9. M. Rieth and W. Schommers, in *What is Life?*, eds. H.-P. Dürr, F.-A. Popp and W. Schommers (World Scientific, New Jersey, London, Singapore, 2002).
10. W. Schommers, in *Quantum Theory and Pictures of Reality*, ed. W. Schommers (Springer-Verlag, Heidelberg, New York, 1989) and *Space and Time, Matter and Mind*, (World Scientific, New Jersey, London, Singapore, 1994).
11. Douglas R. Hofstadter, *GÖDEL, ESCHER, BACH*, Vintage Books, New York, 1980.
12. W. Schommers, *The Visible and the Invisible* (World Scientific, New Jersey, London, Singapore, 1998).
13. W. Schommers, *Quantum Processes* (World Scientific, New Jersey, London, Singapore, 2011).
14. W. Schommers, *Cosmic Secrets: Basic Features of Reality* (World Scientific, New Jersey, London, Singapore, 2012).
15. Wolfram Schommers, *Die Frage nach dem Ganzen*. Weltentwürfe in Physik und Philosophie (Die, Graue Edition, Zug/Scheiz, 2012).
16. Hoimar von Ditfurth, *Der Geist fiel nicht vom Himmel* (Deutscher Taschenbuch Verlag, München, 1980).
17. M. Talbot, *The Holgraphic Universe* (Grafton, London, 1991).

18. M. Talbot, *Mysticism: The New Physics* (Routledge & Keagon Paul, London, Henley, 1981).
19. C. G. Jung, In: J. Ziman, *Reliable Knowledge* (Cambridge University Press, 1978).
20. N. Rescher, *The Limits of Science* (University of California Press, Berkeley, Los Angeles, London, 1984).
21. J. D. Barrow, *The Artful Universe* (Little, Brown and Company, Boston, 1995).
22. D. W. Thompson, *On Growth and Form* (Dover, New York, 1992).
23. K. W. Ford, *The World of Elementary Particles* (Blaisdell, New York, 1963).
24. R. P. Feynman, *The Character of Physical Law* (M.I.T. Press, 1965).
25. D. Ratner and M. Ratner, *Nanotechnology and Homeland Security* (Prentice-Hall International, New Jersey, 2004).
26. M. Rieth and W. Schommers, *Handbook of Theoretical and Computational Nanotechnology* (American Scientific Publishers, Los Angeles, 2006).
27. E. Pastalkova, P. Serrano, D. Pinkhasova, E. Fallace, A. A. Fenton and T. C. Sacktor, *Science* 313, 1141 (2006).
28. N. Kasabov and L. Beniskova, *J. Comput. Theor. Nanosci.* 1, 47 (2004).
29. J. Al-Khalili, *Black Holes, Wormholes and Time Machines* (Institute of Physics, Bristol, 1999).

INDEX

∎ ∎ ∎